C0-BUT-996

CAP-ANALYSIS GENE EXPRESSION
(CAGE)
THE SCIENCE OF DECODING GENE TRANSCRIPTION

CAP-ANALYSIS GENE EXPRESSION
GENE EXPRESSION
(CAGE)
THE SCIENCE OF DECODING GENE TRANSCRIPTION

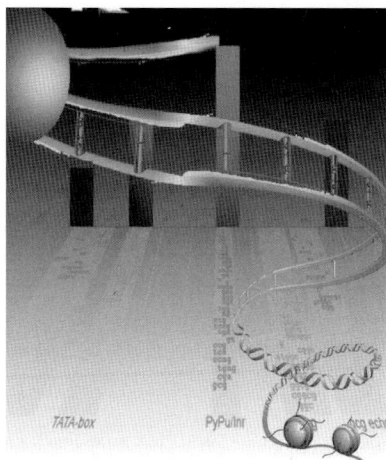

EDITOR

PIERO CARNINCI
RIKEN, Japan

PAN STANFORD PUBLISHING

Published by

Pan Stanford Publishing Pte. Ltd.
Penthouse Level, Suntec Tower 3
8 Temasek Boulevard
Singapore 038988

Email: editorial@panstanford.com
Web: www.panstanford.com

QH
450
.C37
2010

British Library Cataloguing-in-Publication Data
A catalogue record for this book is available from the British Library.

CAP-ANALYSIS GENE EXPRESSION (CAGE)
The Science of Decoding Gene Transcription

Copyright © 2010 by Pan Stanford Publishing Pte. Ltd.

All rights reserved. This book, or parts thereof, may not be reproduced in any form or by any means, electronic or mechanical, including photocopying, recording or any information storage and retrieval system now known or to be invented, without written permission from the Publisher.

For photocopying of material in this volume, please pay a copying fee through the Copyright Clearance Center, Inc., 222 Rosewood Drive, Danvers, MA 01923, USA. In this case permission to photocopy is not required from the publisher.

ISBN-13 978-981-4241-34-2
ISBN-10 981-4241-34-2

Printed in Singapore.

b2751082

Preface

About the time when the draft of the human genome sequence had first appeared, I kept asking myself if there is a systematic, scalable approach to decoding regulatory elements. I also wondered "how can we now understand the network between transcription factors and the genes that they activate and can we have a technology that can be applied to any biological question and sample"?

Even then it was possible to map expressed sequencing tags (ESTs) to the genome, to locate promoter regions and somehow correlate this with expression, but it was too expensive to do it routinely. I had a crazy idea one Sunday afternoon after hours of drawing on a notebook: what if I could chop and concatenate the 5′ ends deriving from full-length cDNA and sequence more than 10 or 20 of these short tags in a single sequencing run? Despite the technical challenges it seemed to make sense: by aligning short tags to the genome, it would become possible to detect all the active promoters and then the expression of transcription factors. I concluded that, if more libraries than the number of existing transcription factors and regulators are made, we could theoretically detect the network between these proteins and the regulated genes. Thus cap-analysis gene expression (CAGE) was conceived and I thought it was worth giving it a try.

After a couple of years of overcoming technical challenges, my team could create a protocol for CAGE with a method of profiling gene expression by producing and sequencing 20-nt short sequencing tags corresponding to the beginning of the RNAs. CAGE technology was readily employed in the Fantom3 project. This revolutionized our understanding of the genome, because we unexpectedly found that the genome produces a much larger variety of RNAs than earlier reported. The genome sequence alone gave us enough information to explain its functions, but with CAGE it became possible to promptly identify novel mRNAs, non-coding RNAs and their promoters, as we did in FANTOM, ENCODE and a growing number of other projects.

These projects have challenged the dogma: contrary to expectations, most genes produce multiple mRNAs and non-coding RNAs starting from multiple promoters. Ultimately, CAGE analysis can elucidate the relationship between the mRNAs and the promoters that control their expression in order to decipher the networks that regulate gene expression and the transcription factors.

Moreover with CAGE, we can comprehensively identify the exact locations of the genome from which the mRNAs originate, identify core promoters, and simultaneously quantify RNA expression levels. Therefore, CAGE has been broadly adopted to infer transcriptional networks because it provides the tools to understand the molecular mechanisms underlying gene expression. CAGE becomes even more valuable when it is used in conjunction with next-generation sequencing technologies, which make CAGE cheaper and more informative than current microarrays.

This book is a guide for current and potential users of CAGE technology who wish to reveal molecular mechanisms in CAGE experiments. The book includes protocols and a guide to the bioinformatics analysis of CAGE datasets, including the design of software and tools for constructing web resources or using existing genome browsers to customize data. I hope that the chapters will be particularly useful to those who are not yet specialists in the field, and provide them with a guide for setting up CAGE technology and/or analysis in their laboratories. This book also provides examples of applications written by the first group of scientists to use CAGE technology for promoter identification, genome annotation, identification of novel RNAs and reconstruction of models of transcriptional control and networks, which may help the readers extracting additional biological insights from the published data.

In conclusion, CAGE technology offers a revolutionary approach for a growing number of scientists beyond early users of genome sequencing centers. This book introduces CAGE technology and its analysis to a broad readership with interests in expression analysis, transcriptional control, marker identification, molecular diagnostics, analysis of networks and RNA biogenesis. I hope the scientists, postdocs, students, technicians and all other readers will expand these approaches to a variety of biological problems using different models and bring forth exciting results to enrich our knowledge of biological systems. Exciting times for scientific discoveries are ahead.

My final thoughts are for the people that have been working with me over years at RIKEN and in the FANTOM consortium and all other collaborators. There are many technicians that have diligently developed experimental conditions and others that have carefully analyzed larger and larger datasets. I am the most grateful to the scientists that have inspired the analysis and interpretation of the CAGE data: their contributions have been essential and the process of discovery has been excitement and fun. Nothing could have been done in isolation, and more excitement lies ahead from the analysis of rich datasets inherent in CAGE libraries.

Piero Carninci

Omics Science Center, RIKEN, Yokohama Institute

Contents

Chapter One

Cap Analysis Gene Expression (CAGE)

Y. Hayashizaki

Omics Science Center, RIKEN Yokohama Institute, Japan
Email: yosihide@gsc.riken.jp

Cap Analysis of Gene Expression (CAGE), originally developed by our group, is used to perform a genome-wide survey of promoters. Our group is primarily engaged in transcriptome analysis and has developed fundamental technologies required for such analysis. In addition, we have developed a full-length cDNA selection method known as the Cap Trapper method (see Chapter 2) that is one of the fundamental technologies for CAGE. In CAGE, the 5'-CAP of mRNA is obtained using this full-length cDNA selection method, or Cap Trapper. From the 5' end of mRNA, which is a product of transcription, 20-base pair are collected as tags using Type IIS restriction enzymes and the tag sequence is determined by sequencing. Incidentally, we have recently developed a technique producing 27-base pairs long tags. The obtained tags can identify promoter sites by mapping model organism genomes to the human. Furthermore, by investigating the expression frequency of the tag mapped to each promoter site, we can measure the transcription activity of promoters. Thus far, CAGE has been the main method for identifying promoters and determining their activity at the genome-wide level.

An analysis of promoters is essential if we are to understand the gene network that comprehensively explains life phenomena at a molecular level. In the international project called Functional Annotation of Mammalian cDNA (FANTOM3), more than 230,000

Cap Analysis Gene Expression (CAGE): The Science of Decoding Gene Transcription **edited by** P Carninci
Copyright © 2010 by Pan Stanford Publishing Pte Ltd
www.panstanford.com
978-981-4241-34-2

mouse promoters were identified by CAGE, and on average that represents approximately 5 promoters for each gene, and therefore the same number of transcription start sites. We are currently engaged in the FANTOM4 project through which we aim to link genes to phenotypic characteristics at the genome level. Since systems biology and gene network analyses require promoter-based analysis, CAGE will become increasingly important in the future.

In recent years, $100,000 genome technology has been accomplished and it is not long before we reach $1,000 genome-sequencing technology. Despite a relatively short read length of 30–100 bases, companies such as 454 (Roche Diagnostics), Solexa (Illumina), SOLiD (Applied Biosystems) and Helicos have rapidly commercialized their own sequencers based on different technologies. As it stands, all of these technologies allow sequencing of a large number of reads: 300,000 to 500,000,000 in a massively parallel way. Often the read length of these technologies is short, so we need to optimize the assembly of trace sequences to determine de novo sequencing of genomic DNA. However, current sequencers are certainly effective for the human genome, where the full-length sequence has already been determined. Furthermore, unlike a traditional cDNA sequencing project, the produced sequence does not identify the correspondence of expressed tags with the original mRNA, and direct use of these data for full-length cDNA sequence reconstruction is difficult. However, these sequencing technologies are very effective for CAGE because it does not require assemblies of full-length cDNA sequences. Since each produced sequence is a tag, all information can be obtained by simply mapping such tags onto the genome. With such massively parallel sequence technologies and the dramatic increase in throughput, it is possible to achieve quantitative analyses of gene expression by measuring the number of tags mapped to each promoter. It is not an exaggeration to say that $1,000 genome sequencing technology is conceivable when combined with CAGE. A massive parallel shotgun sequencer with the ability to produce an enormous number of tags can easily sequence 1,000,000 to 10,000,000 tags per cell in a run. If this is applied to a CAGE library from one tissue, it is theoretically possible to identify and measure RNA molecules at a molecule for each 1 to 10 cells with a probability of greater than 99.9%. Therefore, CAGE-based analysis has entered a completely new phase thanks to these sequencing technologies. In the FANTOM3 research released in 2005, CAGE data was used mainly for genome-wide identification of transcription start sites. Such data provide an extremely useful tool for linking expression analysis to network analysis in the current FANTOM4 analysis on-going at Omics Science Center and at collaborative facilities.

The trend is gradually moving from genome and transcriptome analyses to the molecular description of life phenomena. This may also allow systems biology to systematically cover all molecules inside the living body. In this era, promoter analysis becomes very important and represents an essential analytic target for the investigation of the networks that regulate life phenomena. After our cDNA discovery research (FANTOM1 and 2) using full-length cDNA technology and promoter discovery (FANTOM3) achieved by CAGE, next-generation research in FANTOM4 is about to explain certain cellular conditions with transcription factors. CAGE is one of the key technologies of a pipeline called Life Science Accelerator (LSA) for describing molecular networks that link genes to phenotypes in a computational and combinatorial way.

Both CAGE and SAGE produce fragments of RNAs as tags and determine their sequences. However, the concept behind each is completely different. SAGE was originally developed to obtain frequency information of the number of specific sequences (tags) that appear from the recognition sites of restriction enzymes in the mRNA pool. The one-to-one correspondence between SAGE tags and mRNA/cDNA are thus defined by the sequences of the restriction enzymes used for cDNA cutting and the length of the produced sequences using Type IIs restriction enzymes, like *Mme*I, 20 nt. However, the frequencies of SAGE tags reflect the overall expression frequency of all RNAs containing the tag sequences, regardless of their alternative promoters and splicing. Thus, SAGE does not uncover various promoters of genes and associated transcription activity, nor is it able to measure the number of molecules per RNA (per RNA isoform).

Conversely, CAGE was developed to determine the position of the genome from which RNAs are transcribed. In other words, this technology identifies the core promoter sites of each gene on the genome. Since the tags start from the CAP sites in CAGE, we can identify transcription start sites by mapping the tag to the genome sequence. SAGE can identify the existence of certain expressed sequences, but it cannot differentiate whether the sequences are derived from one or more transcripts or are regulated by various promoters; therefore, SAGE only achieves overall expression information. In this sense, information obtained by SAGE is similar to microarrays. Conversely, information obtained by CAGE indicates the number of molecules produced by each transcription start site. Significantly enough, in CAGE, promoter activity at each alternative "first" exons is identified with the number of the detected tags. In other words, CAGE is the only method that analyzes expression activity at each promoter.

In general we would like to address two major biological questions:

(1) How is production of RNAs regulated?
(2) What kinds of RNAs are produced?

For the first question, we can obtain information by CAGE analysis. For the second, we need synthesis of full-length cDNA and their sequencing. Figure 1.1 shows the advantage of CAGE in measuring regulation by upstream promoters. In a comparison of normal and cancer cells, CAGE is used to show the significant differences in the activity of upstream promoters.

At present CAGE does need 25–50 μg of RNA for quantitative analysis. One of our next targets is to reduce the amount of sample needed for CAGE analysis. Ultimately, we should be able to detect the products of a single cell. This is because conditions of cells vary depending on each cell and if more than one cell is targeted for CAGE measurement, we can only obtain averaged results and thus cannot uncover unique activity at the cellular level.

In the following, we discuss the rapid improvements in the analytical capability of CAGE and explain the principles underlying the technology and associated details of data analysis and applications.

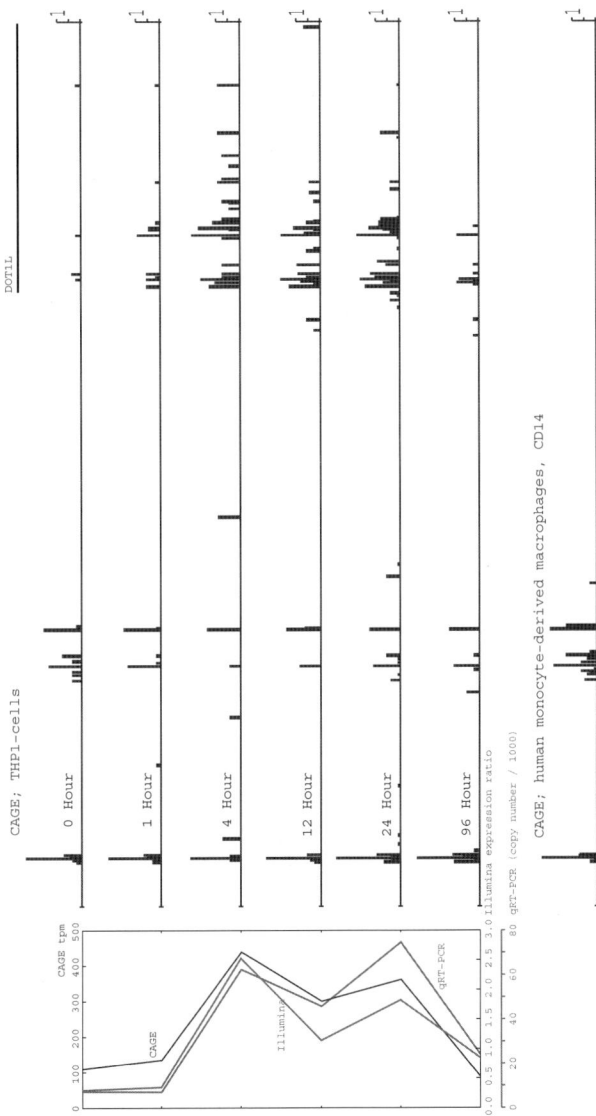

Figure 1.1. Core promoter activity identified by CAGE. CAGE tags are represented by blue vertical bars, mapping at different positions on the genome. CAGE identifies activity of normal tissues (left) and cancer-specific promoters (right). The 6 horizontal lines identify different CAGE libraries from the THP-1 cell line.

Chapter Two

Tagging Transcription Starting Sites with CAGE

Piero Carninci

Omics Science Center, RIKEN Yokohama Institute, Japan
Email: carninci@riken.jp

The output of the genome is much more complex than previously thought. The number of transcription starting sites exceeds by at least one order of magnitude the number of protein-coding genes. CAGE is the only technology to address such transcriptional variability, allowing identification of promoter and the detection of their activity, as well as the detection of novel species of non-protein coding RNAs. CAGE technology provides specific information that cannot be obtained with other transcriptome approaches.

2.1 THE OUTPUT OF THE GENOME IS COMPLEX

In present-day biology textbooks the workings/function of the mammalian genes are/is explained in a fairly simple/elementary and straightforward manner. In a such view, a typical gene transcribes a messenger RNA (mRNA) encoding one or sometimes multiple protein isoforms by alternative splicing. Generally, gene expression is depicted as being driven by a single promoter. Except for splicing, 5′ end capping, and 3′ poly-A tail addition, there are few conceptual differences in the description of prokaryotic and eukaryotic RNA outputs. However, recent studies have challenged this notion of mammalian gene structure and demonstrated that the output of mammalian genomes is much more complex than previously thought. In fact, the complexity of the

Cap Analysis Gene Expression (CAGE): The Science of Decoding Gene Transcription **edited by P Carninci**
Copyright © 2010 by Pan Stanford Publishing Pte Ltd
www.panstanford.com
978-981-4241-34-2

organ and tissue development in the animal kingdom, including the nervous system, is associated with a parallel increased complexity of the genome output in terms of RNA complexity and diversity. This includes a plethora of different RNAs from individual loci, including non-coding RNAs (ncRNAs). Indeed, loci encoding for known proteins also produces multiple transcripts that often originate at different positions in the genome. This includes alternative promoters and promoters inside known gene elements (internal exons, introns as well as 3′ untranslated regions, or 3′ UTRs).[1,2] Moreover, many large non-protein-coding RNAs have also been broadly identified in mammalians,[3,4] and these are often transcribed in a direction opposite to previously known transcripts, otherwise known as antisense (AS) RNAs.[5] The emerging picture, which is still far from complete, is that mammalian genomes can be seen as "RNA machines". This is a concept that is dramatically different than that presented in today's textbooks where the genome is described as a repository of information encoding for proteins. The traditional idea is that proteins derive from the translation of a relatively few species of mRNAs, which are generally produced from a well defined starting site of a single promoter. Instead, this simplistic pattern is true only for a minority of mammalian genes. In fact,

(1) more than half of the genes have alternative promoters;
(2) even defined promoters show multiple transcription initiation sites over a broad span of 50–100 base pairs of core promoters;
(3) an even larger number of genes show alternative splicing;
(4) more than half of the transcriptional units encoded in the genome are constituted by non-coding RNAs;
(5) at least 72% of genes show sense-antisense transcription, often involving non-coding RNAs;
(6) mammalian cells contain a multiplicity of short RNA variants of large RNAs.[6] Such complex patterns of multiple and different RNAs produced by each locus are challenging our image of what constitutes a gene product, and ultimately what constitutes a gene.

Although it is still very informative to analyze the transcriptome with conventional approaches, i.e., using microarrays for simple eukaryotes (like yeast) without dramatically compromising the identification of various overlapping mRNA, but more complex eukaryotes, and in particular vertebrates, show a much large number of transcripts and variants originating from different positions of the same locus. Therefore, predicting all the various RNA products and their transcription starting sites, including the alternative promoters, is impossible without a dedicated approach like CAGE, which has been instrumental in revealing such complexity.[4,7,1]

Indeed, the CAGE technology addresses multiple questions in the transcriptome analysis:

(1) How can we comprehensively identify the transcriptional starting sites (TSSs) and therefore the core promoter elements?

(2) How can we identify the nearby genomic regions that act as core promoter elements?

(3) How can we identify the RNA starting sites that are tissue- and cell-specific?

(4) How can we connect specific transcription starting sites with the regulatory network behind the specific RNA expression in tissues, cells, or other biological conditions?

(5) How can we also identify and profile various capped RNAs, including ncRNAs and antisense RNAs and their controlling elements?

We will address these questions herein and provide more detail in the chapters describing bioinformatic and biological analyses.

2.2 MAPPING 5′ ENDS: FROM ESTS TO TAGGING TECHNOLOGIES

Initial genomics-based approaches for large scale mapping of transcription starting sites (TSSs) have been based on the preparation of full-length cDNA libraries, followed by the sequencing of expressed sequencing tags (ESTs) from their 5′ ends and alignment to the human genome.[8] The strategy underlying this approach is as follows: if multiple 5′ end sequences of full-length cDNAs can be aligned at nearly overlapping positions on the genome, they are likely to identify true TSSs. These initial studies allowed initial mapping of a relatively small number of promoters (only 1031), due to the low throughput and high cost of classic Sanger sequencing (1–2 US dollar per sequence) of individual plasmids.

A technology called Serial Analysis of Gene Expression (SAGE) has been using short fragments deriving from cDNA stretches called "tags" to reduce the cost for each bit of information of such digital expression profiling; tags are then sequenced after concatenation.[9,10] Since concatenation allows cloning of more than 10–20 tags per plasmid, costs are cut down by at least one order of magnitude by sequencing multiple tags per each sequenced plasmid. Using SAGE, typically tens of thousands tags have been routinely achieved in an experiment, grouped, assigned to known mRNAs and genes, and counted as a digital expression profiling. It is important to note that SAGE identifies sequences that are internal to the mRNA, generally towards the 3' end of the transcripts. Therefore, SAGE does identify either 5' or 3' mRNA ends. Moreover, considering that there are at least 300,000 RNAs transcribed by RNA polymerase II in a typical cell, the sequencing depths have been insufficient for detection of rare RNA expression[11] (see also Chapter 15 for an analysis of the deepness of sequencing necessary to analyze a biological sample).

We took advantage of our full-length cDNA isolation technology by combining it with tagging strategies resembling the SAGE approach. After first strand cDNA synthesis and selection of full-length cDNAs by the cap-trapping method,[4,7,1] we have been preparing tags that correspond to the 5' end of mRNAs. We called this technology Cap-Analysis Gene Expression (CAGE), because it allows expression profiling of the 5' ends of the capped RNAs.[12,13] We have also employed the disaccharide trehalose in the first strand cDNA synthesis reaction so we can perform the cDNA synthesis at a higher temperature and therefore maximize synthesis of full-length cDNAs by melting RNA structures and stimulating enzymatic activity.[14,15] After isolation of the 5' cDNA stretches, long tags are excised with a class IIs restriction enzyme, called *Mme*I, which cleaves ∼20 nt into the cDNA and produces tags. More recently we have introduced *Eco*P15I; it cleaves 27 bp into the cDNA end (unpublished results). Tags are PCR-amplified, concatenated. The original protocol included a step of cloning in in plasmids finally followed by sequencing (Fig. 2.1).

A similar technology has appeared more recently, called 5'-SAGE,[16] that uses a series of enzymatic treatments, consisting of uncapped RNA dephosphorylation, followed by decapping and addition of a linker with RNA ligase. This three-step reaction is called "oligo-capping" and is done instead of the chemical reactions that we use for the cap-trapping. The results produced with these two methodologies are conceptually equivalent. However, we have found it easier to work with the cap-trapping technology employed for the CAGE, because it avoids the risk of exposing the RNAs to lengthy enzymatic treatments, with the consequent hazard of undesired accidental RNA cleavage. There are other approaches for preparing tags from mRNAs for various types of profiling.[17]

A remarkable tagging method is the PET (pair-end tag) technology, where ditags are prepared from both the 5' and 3' end of the same full-length cDNA molecules.[18] Di-tagging methods are also extensively described in Chapter 4. Besides identifying TSS and core promoters, these ditags also identify the 3' end (transcription termination sites, TTSs) of the same mRNA molecules. Although this is ideal for defining the borders of transcribed genes, the preparation of ditag libraries has required an intermediate step of amplification of plasmid cDNA libraries before the purification of the 5'-3' ditags. Since plasmids replicate at different rates depending on their size, this may alter the representation of the original RNA in the sequenced library. In fact, plasmids containing long inserts amplify less efficiently than plasmids containing short inserts, and this makes it hard to compare absolute expression of RNAs with different lengths. Future ditag library preparation technologies that do not require cDNA cloning will surely be very beneficial to the transcriptome field.

The technological evolution of CAGE and other tagging technologies has been further influenced by the recent introduction of the 2nd generation of sequencers (454 Life Sciences, Solexa/Illumina, SOLiD and other upcoming sequencers; see also Chapters 1 and 5). Indeed, sequencing for digital expression measurements has become progressively cheaper, which is very welcome news for genomics and other fields. The 454 Life Sciences sequencer still uses concatenation to take full advantage of the long runs; however, concatenation is no longer necessary for other platforms that produce several to hundreds of millions of short sequencing reads. This adds further simplification to the preparation of CAGE libraries and brings the cost of sequencing technologies closer to the level of microarrays (see also Chapter 5 for a discussion on sequencing platforms).

2.3 LINKING CORE PROMOTERS TO GENOMIC ELEMENTS

Tagging technologies based on isolating and profiling cDNA ends are clearly needed to finely identify the borders of RNAs, whereas CAGE particularly focuses on the identification and profiling of 5′ capped ends and therefore core promoter elements. However, these technologies do not identify long-range regulatory

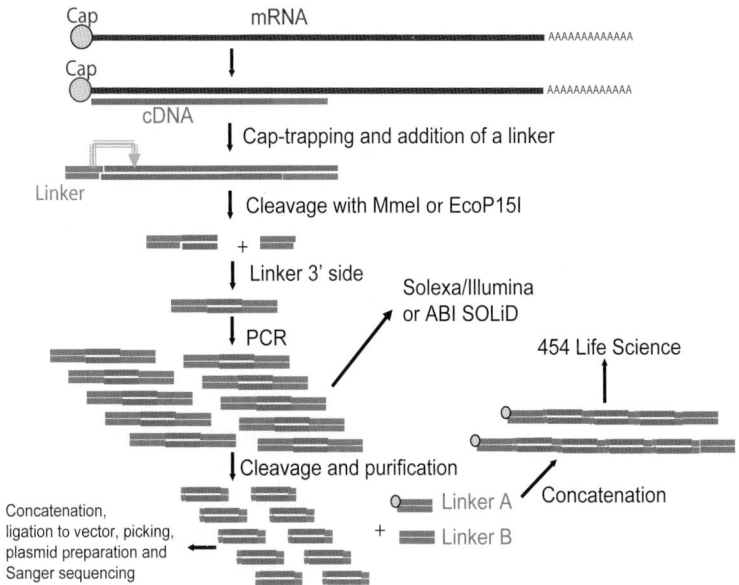

Figure 2.1. Workflow in preparation of CAGE libraries. The cDNA is randomly primed or alternative oligo-dT primed or primed with a combination of both methods (see also Chapter 3). Subsequent to the first strand cDNA synthesis, only the capped RNAs, where the cDNAs (red) have reached the cap-site, are isolated with cap-trapper, followed by addition of a linker to the single-strand cDNA and second strand cDNA synthesis. The linker at the 5′ end is recognized by a class IIs restriction enzyme, *Mme*I or *Eco*P15I, that cleaves 20 or 27 nt, respectively, into the cDNA. After the removal of the remainder of the cDNA and addition of other linkers, the CAGE tags are amplified and used for direct sequencing with SOLiD or Solexa. Or, after cleavage of the external part of the linkers, they are concatenated for classic Sanger sequencing (nowadays not in use for CAGE) or the 454 Life Sciences sequencer.

elements like enhancers. This will require new technology. Chromatin immunoprecipitation followed by whole genome tiling arrays, or more recently by high-throughput sequencing, indeed allows identification of regulatory elements bound to transcription factors and other DNA-binding proteins. However, new technologies will be required to link long-range regulatory elements capable of mapping hundreds of Kbp away from the gene they control and the core promoters under their influence. Since at least some of the long-range controlling elements and controlled RNAs are known to be co-localized in the nucleus, technologies evolving from the Chromosome Capture Conformation ("3C" technology) will be important for integrating these long-range regulatory elements in regulatory networks (see also Chapter 4). In fact, 3C captures the interaction of chromatin regions.[19]

Another area is the identification of all the set of the exons that are linked to specific CAGE tags. In fact, alternative promoters are often linked to alternative splicing patterns, where certain exons are specifically used in association with sets of promoters.[20] There are various potential solutions to this problem. The best established, but prohibitively costly and time consuming, is the extensive sequencing of full-length cDNA clones. Alternatively, promoter-exons association derived from novel algorithms will allow the assembly of shotgun transcriptome libraries and the identification of exact splicing patterns. At the moment, we envisage that core promoter-exon association will be achieved by sequencing both ends of random-primed cap-selected cDNAs. Yet another alternative will arrive when high-throughput sequencing of long, single molecules becomes possible.[21] Direct full-length sequencing of full length cDNAs should also have an important role in transcript structure identification and, if unbiased for frequency, expression profiling.

2.4 CDNA ENDS OR THE WHOLE SEQUENCE?

How do RNA tag-profiling approaches (like CAGE and 5'-3' ditags) compare with whole transcriptome profiling, such as the whole genome tiling arrays interrogation of mRNAs[3] or whole transcriptome shotgun,[22,23] all of which integrate the transcribed sequences in a single experiment? Whole transcriptome technologies have indeed shown that a large fraction of mammalian genome is expressed, including the expression of non-annotated loci and non-coding RNAs. However, identification of transcribed DNA strand has not always been possible. Additionally, even if the orientation of the identified transcript can be identified,[22] it has not been possible to unambiguously identify the origin and termination sites and the connectivity between all exons, in particular for genomic regions producing multiple RNAs. This is especially so because of frequent overlaps of multiple transcripts.[4] In mammalians, the majority of the genes have alternative promoters and there are capped RNA molecules mapping in the middle of exon sequences and more frequently the 3' UTR regions in the last exon (3' UTR transcripts). In fact, these 3' UTR transcripts have been found to map to, and likely originate from, conserved regions in the 3' UTRs. Genomic regions around and downstream of these 3' UTR transcripts are often conserved and promote transcription in a reporter assay.[1] Other transcripts have been found that overlap the promoters (PASRs, promoter associated short RNAs) and termination sites (TASRs, or termini associated short RNAs),[2] and this further complicates the assembly of individual transcripts from whole transcriptome shotgun experiments. In contrast, CAGE profiling clearly identifies 5' edges of such capped molecules and putative core promoter elements.

If we compare CAGE with whole transcriptome shotgun and microarray expression profiling, there are key points to bear in mind. In several microarray platforms, the probes are designed to match at the 3' ends (3' UTRs) of mRNAs. Since there are many independent, putative non-coding RNAs that originate within genes and span the 3' UTRs regions, we argue that it is inaccurate to profile mRNAs expression with 3'-end probes and relate these data to the frequency of the entire protein-coding mRNAs. For instance, signals from 3' UTR transcripts may also derive from non-coding RNAs that overlap known protein-coding mRNAs, confounding the signal from the true mRNAs. In addition, RT-PCR expression measurements require careful design of the regions to be tested and multiple sets of primers are ideally used. In contrast, CAGE addresses these issues by identifying all capped molecules so that one can decide to use only the information deriving from the core promoters upstream of the mRNA coding regions and also to separate the contribution of individual core promoters.

Ultimately, once the mRNA maps are completed for all cells in many model organisms, we will be able to assign CAGE tags to

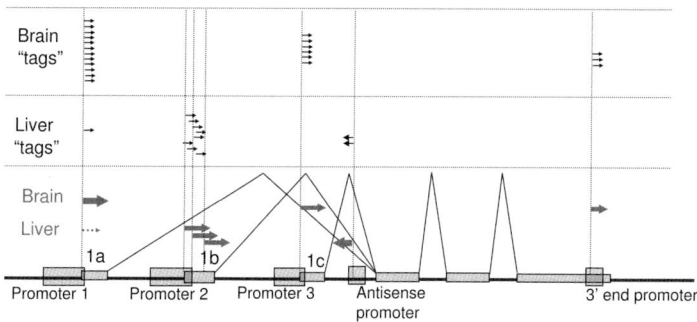

Figure 2.2. Identification of alternative core promoters (pink shadowed boxes) in a hypothetical gene model, characterized with 3 "classic" promoters that transcribe three main mRNAs isoforms including different first exons (1(a), 1(b), and 1(c)). The preparation of CAGE libraries from "brain" and "liver" produce tags (black arrows) that in turn identify the TSS originated by different promoters (1, 2, and 3). The transcription consensus for each library is represented by red arrows. Promoters can have a different shape based on the distribution of tags: sharp distribution, where most tags map to a unique position (1 and 3), or broad (RNA originates from multiple starting sites in a broad area).

mRNA and ncRNA variants and finely define the expression for each gene product from each different promoter. Figure 2.2 shows an example of assignment of CAGE tags to different transcripts and promoters.

Additionally, CAGE tags very accurately identify the transcribed genomic DNA strand for detection of sense/antisense transcription. For this, CAGE tags have been invaluable to the study of sense-antisense transcription,[5] identifying that at least 72% of the mammalian transcriptional units show sense-antisense transcription. Moreover, bidirectional promoters have been identified as well, showing frequent patterns of transcription in a direction that is antisense to the main known transcript, starting from the position upstream of the main mRNA transcriptions starting site.[24]

2.5 IDENTIFICATION OF FUNCTIONAL ELEMENTS IN THE GENOME

Do the CAGE tags truly identify promoters and regulatory elements? Extensive validation based on rapid amplification of cDNA ends (RACE) has proven the existence of capped RNA molecules in correspondence with the majority of the tested tag cluster (>90%), including the ones showing low expression. We define CAGE clusters expression with an absolute number: we count the number of tags and normalize them to one million tags, hereafter defined as unit of expression in transcripts per million, or TPM. Accordingly, we have successfully validated most of the TSS expressed at 10 TPM or above. For highly expressed tags, we have almost always identified RACE products. CAGE tag clusters that are not verified by RACE may be the result of false positive signals or they fail because of problems in the RACE primer design (the exons are not always available). Furthermore, since RACE experiments have often not covered all the tissues initially used to prepare CAGE libraries, we think that the large majority of the CAGE signal identifies capped RNAs. CAGE identifies capped molecules in the majority of tested targets, including 3' UTR promoters, exons and introns and antisense coding/noncoding RNAs.[4] The tags at the 5' of the genes clearly recognize known types of promoters. They often originate from CpG islands and less frequently have a TATA-box.[1] See also Chapter 14 for a comprehensive discussion of the features of promoters. Additionally, clusters of CAGE tags also overlap with transcriptional factors binding sites position identified by Chromatin Immuno-precipitation (ChIP), followed by tiling arrays hybridization or high-throughput sequencing analysis. Some aspects of this analysis are described in the Encode project,[25] for which CAGE tags have been used in the identification of novel transcribed RNAs, their borders, and the map of the core promoters and regulatory elements. There is generally good correlation between CAGE, ChIP, and DNase hypersensitive sites in the genome, particularly for the highly active promoters.

Understanding the mRNA isoforms expression from specific promoter of is the key to deciphering transcriptional networks. In fact, since different core promoters are found to be responsible for the transcription of different mRNAs involved in various biological processes, it is now possible to identify with CAGE the specific set of promoters and their activating transcription factors that are controlling specific core promoters (see also Chapters 11 and 12). Thus, transcription factor binding on promoters and the regulatory networks can be predicted. Subsequently, such predicted networks can be validated by CIP to verify that predicted transcription factors do actually bind predicted targets. Additionally, knocking down transcription factors, followed by CAGE analysis, provides an alternative approach to a fine mapping of both global effects, such as the regulation of expression of mRNA species, and local effects, such as shifts in transcription starting site usage within core promoters (personal communication from Alistair Forrest, unpublished). This observation agrees with earlier observations showing that transcription starting site usage varies within the same set of core promoters in different tissues.[26]

2.6 TECHNOLOGY EVOLUTION, SAME LESSONS?

With the arrival of the third-generation sequencers (based on single molecule sequencing), it is important to further improve global detection of 5′ end-based expression profiling. For instance, there is a need to prepare longer reads to assign unique genomic regions to about 20% of CAGE tags that currently map on multiple positions (see also Chapter 8 for a strategy on how to rescue tags that map on multiple genomic locations). Another way is to read long ditags from 5′ and 3′ of cDNA molecules in order to assign CAGE tags to the genes and genomic regions. Ideally we want to simply map full-length cDNAs onto the genome after direct sequencing on these platforms as individual cDNAs. However, even though the switch to long reads for exploring the structure of the RNAs will be useful, we believe that tagging technologies present the best choice for quantitative expression profiling and detection of the difficult-to-trace targets such as very long mRNAs. Indeed, capturing and amplifying short DNA tags is less severely associated with size bias during the amplification of the targets before sequencing. Heat denaturation and primer extension of 20-27 nt insert amplification is likely to be much better than handling and amplifying full-length cDNA molecules of heterogeneous sizes. Direct sequence of 5′ and 3′ ends of full-length cDNAs is unlikely on existing platforms. In fact, the bridge PCR step is needed in Illumina/Solexa sequencing of the DNA molecules attached to the sequencing slides to amplify the target molecule before sequencing reactions. Bridge PCR is limited to inserts up to 1 Kb. The ABI SOLiD and the 454 Life Science sequencing platforms require somewhat equivalent amplification steps, called "emulsion PCR", and these are limited to fragments up to about 1 Kb. In both bridge

PCR and emulsion PCR, the reaction yields are extremely variable for DNA molecules of different CG content and length. Both the Solexa and the SOLiD platform perform sequencing from both ends, and it is tempting to read the cDNA to produce 5′-3′ ditags. However, a large part of full-length cDNAs are longer than 1 Kb, and they can be even longer than 10 Kb; all of these would be indeed a very problematic substrate for unbiased amplification, because a large fraction of full-length cDNAs would be missing from the sequencing output. Several companies such as Pacific Bioscience (http://www.pacificbiosciences.com), are developing instruments capable of sequencing long DNA molecules, and these instruments may be suitable for full-length cDNA sequencing. If size-unbiased, cap-trapped full-length cDNA molecules could be entirely sequenced, this would be revolutionary for the simultaneous mapping of promoters, alternative splicing, and termination sites. For now, the best option is to identify the mRNAs associated with CAGE tags by taking advantage of full-length cDNA collections, whole transcriptome shotgun, tiling arrays, and experimental validation. We are also confident that the increased read length of second-generation sequencers using long 5′ and 3′ tags, together with improved algorithms, will help in the assembly of transcripts without full-length single molecule sequencing.

While whole transcriptome shotgun approaches produce data that are often difficult to interpret, CAGE library preparation has become straightforward and CAGE analysis has led to pinpointing and fine analysis of promoter elements. The CAGE protocol is also being standardized to improve its reliability and reproducibility and to simplify comparison between experiments. This is feasible because concatenation is not necessary in the new sequencing platforms of Solexa, SOLiD, and Helicos. In the meantime CAGE protocols will be simplified to prepare multiple CAGE libraries in parallel in few days in a 96-well format in order to perform multiple replicas and simultaneously test them under different experimental conditions. Additionally, we plan to apply the same lessons we learn from large-scale CAGE to the miniaturization of CAGE technology (nanoCAGE; Plessy *et al.*, manuscript in preparation) so that we can target small subsets of specialized cells such as individual neuron populations, developmental stages, stem cells, and even RNAs from subcellular compartments isolated from large tissues.

ACKNOWLEDGMENTS

This work was founded by a Research Grant for RIKEN Omics Science Center from MEXT, a grant of the 6th Framework of the EU commission to the NFG consortium, and a Grant-in-Aids for Scientific Research (A) No.20241047 for PC.

References

[1] P. Carninci *et al.* Genome-wide analysis of mammalian promoter architecture and evolution. *Nat. Genet.* (2006).

[2] P. Kapranov, *et al.* RNA maps reveal new RNA classes and a possible function for pervasive transcription. *Science* **316**, 1484–1488 (2007).

[3] J. Cheng *et al.* Transcriptional maps of 10 human chromosomes at 5-nucleotide resolution. *Science* **308**, 1149–1154 (2005).

[4] Carninci, P. *et al.* The transcriptional landscape of the mammalian genome. *Science* **309**, 1559–1563 (2005).

[5] S. Katayama *et al.* Antisense transcription in the mammalian transcriptome. *Science* **309**, 1564–1566 (2005).

[6] P. Carninci, J. Yasuda & Y. Hayashizaki. Multifaceted mammalian transcriptome. *Curr. Opin. Cell. Biol.* **20**, 274–280 (2008).

[7] A. Sandelin *et al.* Mammalian RNA polymerase II core promoters: insights from genome-wide studies. *Nat. Rev. Genet.* **8**, 424–436 (2007).

[8] Y. Suzuki *et al.* Identification and characterization of the potential promoter regions of 1031 kinds of human genes. *Genome Res.* **11**, 677–684 (2001).

[9] V. E. Velculescu, L. Zhang, B. Vogelstein and K. W. Kinzler. Serial analysis of gene expression. *Science.* **270**, 484–487 (1995).

[10] S. Saha *et al.* Using the transcriptome to annotate the genome. *Nat. Biotechnol.* **20**, 508–512 (2002).

[11] P. Carninci, Constructing the landscape of the mammalian transcriptome. *J. Exp. Biol.* **210**, 1497–1506 (2007).

[12] T. Shiraki *et al.* Cap analysis gene expression for high-throughput analysis of transcriptional starting point and identification of promoter usage. *Proc. Natl. Acad. Sci. USA.* **100**, 1577–1581 (2003).

[13] R. Kodzius *et al.* CAGE: Cap analysis of gene expression. *Nat. Methods.* **3**, 211–222 (2006).

[14] P. Carninci *et al.* Thermostabilization and thermoactivation of thermolabile enzymes by trehalose and its application for the synthesis of full length cDNA. *Proc. Natl. Acad. Sci. USA* **95**, 520–524 (1998).

[15] P. Carninci, T. Shiraki, Y. Mizuno, M. Muramatsu & Y. Hayashizaki, Extra-long first-strand cDNA synthesis. *Biotechniques*. **32**, 984–985 (2002).

[16] S. Hashimoto *et al*. 5′-end SAGE for the analysis of transcriptional start sites. *Nat. Biotechnol*. **22**, 1146–1149 (2004).

[17] M. Harbers and P. Carninci. Tag-based approaches for transcriptome research and genome annotation. *Nat. Methods* **2**, 495–502 (2005).

[18] P. Ng *et al*. Gene identification signature (GIS) analysis for transcriptome characterization and genome annotation. *Nat. Methods* **2**, 105–111 (2005).

[19] M. Simonis, J. Kooren and W. de Laat. An evaluation of 3C-based methods to capture DNA interactions. *Nat. Methods* **4**, 895–901 (2007).

[20] M. Zavolan *et al*. Impact of alternative initiation, splicing, and termination on the diversity of the mRNA transcripts encoded by the mouse transcriptome. *Genome Res*. **13**, 1290–1300 (2003).

[21] J. Korlach *et al*. Long, processive enzymatic DNA synthesis using 100% dye-labeled terminal phosphate-linked nucleotides. *Nucleotides Nucleic Acids* **27**, 1072–83 (2008).

[22] N. Cloonan *et al*. Stem cell transcriptome profiling via massive-scale mRNA sequencing. *Nat. Methods* **5**, 613–9 (2008).

[23] A. Mortazavi, B. A. Williams, K. McCue, L. Schaeffer and B. Wold. Mapping and quantifying mammalian transcriptomes by RNA-Seq. *Nat. Methods* **5**, 621–8 (2008).

[24] P. G. Engstrom *et al*. Complex Loci in human and mouse genomes. *PLoS Genet*. **2**, e47 (2006).

[25] E. Birney *et al*. Identification and analysis of functional elements in 1% of the human genome by the ENCODE pilot project. *Nature* **447**, 799–816 (2007).

[26] H. Kawaji *et al*. Dynamic usage of transcription start sites within core promoters. *Genome Biol*. **7**, R118 (2006).

Chapter Three

Construction of CAGE Libraries

Hiromi Nishiyori and Piero Carninci[*,†]

Omics Science Center, RIKEN Yokohama Institute, Japan
Email: [†]carninci@riken.jp

This chapter updates an established CAGE protocol[1] to work with the sequencer GS20/GSFLX system (454 Life Sciences).[2] Based on the same protocol, CAGE will work with the Illumina GA or GA2 or the SOLiD sequencing instruments, provided that the user changes the the linkers and adaptors to fit the needs of other platforms. Likewise, these modified protocols will also be useful with third-generation sequencing instruments (singl molecule; see also Chapters 1 and 4) that should perform deep sequencing of each run.

3.1 INTRODUCTION

CAGE tags are derived from the 5′ end of RNA expressed in the cells and tissues. CAGE allows high-throughout gene expression analysis and the simultaneous profiling of transcriptional start site usage and identification of core promoters. It is essential to obtain a large number of CAGE tags to statistically analyze the identified promoters and the expression profile of the majority of the RNA transcripts, including the rarely expressed ones. In doing this it is mandatory to keep the number of molecules large enough to avoid introducing "molecular bottlenecks", which would result in multiple sequencing of CAGE tags derived from the same original RNAs.

---

[*]Corresponding Author

Cap Analysis Gene Expression (CAGE): The Science of Decoding Gene Transcription edited by P Carninci
Copyright © 2010 by Pan Stanford Publishing Pte Ltd
www.panstanford.com
978-981-4241-34-2

We describe here a proven protocol for the CAGE library preparation, which we use starting from at least 30 μg of total RNA, and typically with 50 μg of RNA. This is based on the cap-trapper technology[3] to isolate the most terminal (5′ end) of the cDNAs, corresponding to the cap site of the RNA transcribed by the RNA polymerase II (Pol 2). After selection of the 5′ end, the second strand cDNA is synthesized. Subsequently, we cleave a tag at the 5′ end (20 nt in length; note that the novel enzyme, like *EcoP*15I, will in the future cleave 27 nt apart from the cap-site). Then, 5′ ends tags are isolated and after addition of a second linker, tags are PCR amplified. Finally, after the cleavage, sequences tags are purified and concatenated together with appropriate linkers for sequencing with the 454 Life Science sequencer. New sequencing platforms such as Illumina and SOLiD will use the tags directly, without concatenation.

Please review the list of solutions and reagents listed at the end of this chapter before proceeding with the CAGE library preparation.

3.2 STAGE 1: SYNTHESIS OF FIRST-STRAND CDNA

The first strand cDNA can be primed with random primer (N20) or oligo-dT primer ($T_{14}VN$). In order to maximize the detection of non-polyadenylated RNAs, including ncRNAs, random primer is recommended. Conversely, oligo-dT priming can better identify transcripts that originate from the 3′ UTR[4], which are likely to be short poly-A plus RNAs that are not primed efficiency with random primers. Depending on the purpose of the experiment, we have also used a 1:5 mixture of oligo-dT and random primers.

3.2.1 Synthesis of First-Strand cDNA

Before proceeding, the RNA should be concentrated. Therefore, proceed to step 2 if the RNA is already concentrated.

(1) To 50 μg of total RNA, add NaCl to a final concentration of 0.25 M with an equal volume of isopropanol. Precipitate the RNA at −20°C for 30 min and centrifuge at 15,000 g for 20 min. Wash the pellet twice with 80% ethanol and dissolve the pellet in 20 μL of water. Transfer the RNA solution to a

0.6 ml tube (tube A) and add 2 μL of 6 μg/μL 1st primer ready for the first strand DNA synthesis.

(2) In another tube (tube B), prepare a reaction mixture without RNA.

2× GCI LA Taq buffer (TaKaRa)	75 μL
10 mM each dNTPs	4 μL
Sorbitol/Trehalose mix	30 μL
Water*	4 μL
Reverse Transcriptase, RNaseH minus (200 U/μL)	15 μL
Total volume	**128 μL**

(3) Heat tube A to 65°C for 5 min and then place immediately on ice.
(4) Quickly transfer the contents of tube B into tube A.
(5) Carry out reverse transcription in a thermal cycler as follows: 30 sec at 25°C, 30 min at 42°C, 10 min at 50°C, and 10 min at 56°C; hold at 4°C until further processing. If you are using an oligo-dT primer only as the 1st strand cDNA primer, you can skip the 30 sec at 25°C as priming takes place at 37°C.

*Water is intended here and later as MilliQ, RNase/DNase free water, or equivalent grade deionized nuclease free water.

3.2.2 CTAB/urea Purification

Cethyl trimethyl ammonium bromide (CTAB) is a cationic detergent that with controlled NaCl concentration (0.3 – 0.5 M) will specifically precipitate long nucleic acids by creating specific complexes left in solution with other molecules like proteins, short oligonucleotides and small molecules. Subsequent to the precipitation, CTAB has to be removed by resuspending the pellet in high salt buffer (like NaCl or Guanidine-Cl), followed by final ethanol precipitation.

(1) Transfer the cDNA/RNA hybrids to a 1.5 ml tube and add 2 μL of 0.5 M EDTA and 3 μL of Proteinase K. Incubate the reaction mixture for 15 min at 45°C.
(2) Add 300 μL of CTAB/urea solution and 30 μL of 5 M NaCl. After leaving for 10 min at room temperature, centrifuge at 15,000 g for 15 min. Carefully remove the supernatant and

place it in another tube. This will prevent you from losing your sample.

(3) Dissolve the pellet in 200 μL of 7M Guanidine-Cl. Then add 500 μL of 99.5% ethanol. Precipitate the cDNA/RNA hybrids at $-20°$C for 30 min and centrifuge at 15,000 g for 20 min at 4°C. Wash the pellet twice with 80% ethanol and completely dissolve the pellet in 46 μL of water.

This may be a convenient stopping point (samples can be stored at $-20°$C). The procedure thus far can be stopped at any point during ethanol precipitation except during oxidation and biotinylation, when it is advisable to control the extent of reaction by limiting the time.

3.3 STAGE 2: OXIDATION/BIOTINYLATION

Stages 2 to 4 describe the cap-trapper method in order to isolate cDNAs that have been fully synthesized to the cap site. In Stage 2, biotin is added to the cap-structure.[4] This reaction consists of oxidation of the diol group at the 5′ end and subsequent coupling with the biotin hydrazide. Long-arm biotin hydrazide is preferable in order to more efficiently bind the cDNA/RNA hybrid to the magnetic beads.

3.3.1 Oxidation

(1) Add 3.3 μL of 1M Sodium Acetate (pH 4.5) and 2 μL of 250 mM NaIO$_4$. Keep the tube for 45 min on ice in the dark (by wrapping or covering with an aluminum foil).

(2) Stop the reaction by adding 1 μL of 80% glycerol and mix rapidly. Add 0.5 μL of 10% SDS, 11 μL of 5 M NaCl, and 61 μL of isopropanol. Precipitate the cDNA/RNA hybrids like in stage 1, step 1 and completely dissolve the pellet in 50 μL of water

3.3.2 Biotinylation

(1) Add 5 μL of 1M Sodium citrate (pH 6.1), 5 μL of 10% SDS, and 150 μL of 10 mM biotin (long arm) hydrazide. Mix gently and keep the reaction mixture overnight (10 to 16 hours) at room temperature. (Alternatively we have also tested biotinylation

at 37°C for 3 hours, but some batches of biotin hydrazide have degraded the RNA/cDNA at this temperature.)

(2) Add 75 μL of 1M Sodium Acetate (pH 6.1), 5 μL of 5 M NaCl, and 750 μL of 99.5% ethanol. Precipitate the cDNA/RNA hybrids like in stage 1, step 1 and then completely dissolve the pellet in 180 μL of 0.1 × TE.

At the end of this step, the RNAs (which form hybrids with the first strand cDNAs) carry biotin groups on the cap-site (and at their 3′ ends). Subsequently, RNase I treatment digests any single-strand RNA. The RNAs that are not digested are those regions that are protected by the presence of the first-strand cDNA. Only when the cDNAs extend until the 5′ end (the cap-site), the biotinylated cap is retained on the cDNA/RNA hybrid, and can be used for subsequent purification of 5′-complete cDNA molecules.

3.3.3 RNase I Treatment Removal of Biotinylated Cap when cDNAs do not Reach the 5′ end

(1) Add 20 μL of 10 × RNaseI buffer and RNase I to the final concentration of 1 U/μg of original starting RNA. Incubate the mixture for 30 min at 37°C. If using random primer for the first-strand cDNA synthesis, heat up the tubes at 65°C for 5 min after incubation. This step removes any incompletely synthesized cDNAs that did not extend to the cap site.

(2) Add 4 μL of 10% SDS and 2 μL of Proteinase K (10 μg/μL). Incubate the mixture for 30 min at 45°C.

(3) Extract the cDNA/RNA hybrids with phenol-chloroform (1:1; vol/vol); back extract using 50 μL of water and chloroform. Add 1 μL of 20 μg/μL tRNA, 13 μL of 5 M NaCl and 250 μL of isopropanol. Precipitate the cDNA/RNA hybrids like in stage 1, step 1 and completely dissolve the pellet in 50 μL of 0.1 × TE.

3.4 STAGE 3: CAPTURE-RELEASE

The 5′-complete cDNA/RNA hybrids are now ready to be isolated with Streptavidin-coated magnetic beads, while the cDNAs that did not reach the cap-site are eliminated during the following steps.

3.4.1 Capture and Subsequent Release of 5'-Completed cDNAs

(1) Prepare the Streptavidin-coated magnetic beads (Dynabeads-MP-280; other suppliers may also be suitable). Add 5 μL from a stock of 20 μg/μL tRNA into 500 μL of Streptavidin beads. Incubate on ice for 30 min with occasional shaking. Separate the beads with a magnetic stand for 2-3 min and remove the supernatant. Wash the beads 3 times with 500 μL of wash buffer (4.5 M NaCl/50 mM EDTA). Resuspend the beads with 500 μL of wash buffer.

(2) Transfer 350 μL of the washed beads into the tube containing cDNA/RNA hybrids. After mixing gently, rotate the tube for 10 min at 50°C. Then, transfer the remaining 150 μL of pre-washed beads into the tube containing cDNA/RNA and the first aliquot of beads. After mixing gently, rotate the tube for 20 min at 50°C, separate the beads with a magnetic stand for 2-3 min and transfer the supernatant into a new tube.

(3) Wash the beads gently multiple times with the following buffers. The volume of wash buffer is 500 μL each.

   (a) Two times with 4.5 M NaCl/50 mM EDTA
   (b) Once with 0.3M NaCl/1 mM EDTA
   (c) Three times with 0.4% SDS/0.5 M NaOAc/20 mM Tris-HCl pH 8.5/1 mM EDTA
   (d) Two times with 0.5 M NaOAc/10 mM Tris-HCl pH 8.5/1mM EDTA

   Washing removes all the incompletely synthesized cDNAs from the surface of the beads.

(4) Add 100 μL of release buffer (50 mM NaOH/5 mM EDTA) to the beads. Stir briefly and incubate the beads and release buffer for 5 min at room temperature with occasional shaking.

(5) Separate the beads with a magnetic stand for 2–3 min and transfer the supernatant containing the released cDNA into a new siliconized tube on ice.

(6) Repeat the release cycle 2 more times with release buffer.

(7) Add 100 μL of 1M Tris-HCl (pH 7.0) to the solution containing the released cDNA on ice, mix quickly and quickly place on ice again. Add 1 μL of 10 U/ul RNase I, mix quickly, and

quickly place at 37°C. Incubate the reaction mixture for 10 min at 37°C.

(8) Add 7 μL of 10% SDS and 2 μL of Proteinase K. Incubate for 15 min at 45°C.

(9) Extract the cDNA with phenol-chloroform (1:1; vol/vol); back extract using 50 μL of 0.1 × TE buffer. Add 3 μL of 1 μg/μL glycogen, 22.5 μL of 5 M NaCl and 450 μL of isopropanol. Precipitate the cDNA as for stage 1, step 1 and completely dissolve the pellet in 50 μL of 0.1 × TE buffer.

At this point, the samples should be purified in a spin column to eliminate the shorter fragments (from RNA) and possible residual primers. S400 MicroSpin Columns are commonly used.

(10) Resuspend the S400 resin in the column by vortexing. Loosen the cap and snap off the bottom closure. Place the column in a 2 ml tube and pre-spin it at 735 g for 1 min.

(11) Place the column in a new 1.5 ml tube, slowly apply 50 μL of cDNA to the top-center of resin, then add 50 μL of 0.1 × TE.

(12) Spin the column at 735 g for 2 min. The purified cDNA is collected in the bottom of the tube.

(13) Add 5 μL of 5 M NaCl and 100 μL of isopropanol. Precipitate the cDNA like in stage 1, step 1 and completely dissolve the pellet in 5 μL of water.

3.5 STAGE 4: SINGLE STRAND LINKER LIGATION

The next step requires the addition of a linker to prime the second-strand cDNA synthesis from the cDNA ends corresponding to the cap-sites. There are various choices in terms of sequences of this linker, as we can introduce either a simple second standard primer or primer sets having extra nucleotides, called barcode tags (Appendix).

These tissue/sample barcode tags are used if cDNAs are subsequently pooled. By pooling, all the CAGE tags are subjected at the same PCR condition in the same reaction, minimizing differences among samples due to PCR.[5]

Although the protocol presented here is adapted to the GS20/GSFLX system (454 Life Sciences), different linker design can be used to adapt the CAGE protocol to the other 2nd and 3rd generation sequencers by changing linkers.

3.5.1 Single Strand Linker Ligation

(1) After redissolving the cDNA in $5\,\mu$L ($\sim$1 μg/tube), incubate it at 65°C for 5 min and quicly place on ice. Incubate 2.5 μL of the 0.08 μg/μL 5′ linker at 37°C for 5 min and quickly place on ice to ensure the linkers are annealed.

(2) Add 2.5 μL of 5′ linker, 7.5 μL of solution I and 15 μL of solution II of TaKaRa DNA Ligation Kit, version 2.1. Incubate at 16°C overnight (alternatively, other products using PEG to enhance ligation should also work).

(3) Add 10 μL of 0.1 × TE buffer, 1 μL of 0.5 M EDTA, 1 μL of 10% SDS and 1 μL of Proteinase K. Incubate for 15 min at 45°C.

This is a convenient stopping point; in this case, samples should be stored at −20°C.

If barcode IDs are used, this is when the various cDNAs are mixed. In this case, to evaluate the concentration of the cDNAs in each sample, you can test the efficiency of cDNA amplification by quantitative RT-PCR for specific known genes or by simply checking the yield of second-strand cDNA using small aliquots of the ligation reactions. If samples are bar-coded and mixed, use aliquots of at least 10 μL for each sample and pool together in one tube. If the volume exceeds 40–50 μL, volume adjustment and further purification (ethanol precipitation) are necessary to keep the volume reasonably low for later spin column purification.

(4) Extract the cDNA with phenol-chloroform (1:1) and chloroform. To maximize yield, we additionally perform a second extraction (called back extraction) by adding 60 μL of spin column buffer and recovering the remaining cDNA from the interphase between the water and organic phases. The 1st and 2nd extractions are kept in different tubes. Typically the first extraction is about 40 μL and the second extraction is about 60 μL.

3.5.2 S400 Spin Column

This step is necessary to eliminate the remaining linkers and linker dimers. The length of linker dimers is 106-116 bp (sizes for linkers without/with barcode tags in the current design). Calibration of the column is important to avoid linker contamination on the CAGE libraries, which should only be the minimum contaminant of a CAGE library (e.g: <1%)

(1) Add 2.5 ml of S400 HR matrix into the open tip of the spin column.
(2) Equalize the S400 spin column 3 times with 2 ml of the spin column buffer, each time allowing the column to drain by gravity.
(3) Transfer the column to a 15 ml centrifuge tube and centrifuge in a swing-out rotor at 400 g for 2 min at room temperature.
(4) Test the column by applying 100 μL of the spin column buffer. If the amount is fairly different from the input, repeat this step until the eluted volume is approximately the same as the added one. Equilibrated columns have to be used within a short time (e.g., one hour).
(5) Place a 1.5 ml tube into the 15 ml centrifuge tube to collect the flow through. Then apply the sample into the column and centrifuge at 400 g for 2 min at room temperature.
(6) Collect the flow through in another tube. Then apply 50 μL of spin column buffer on the column and centrifuge.
(7) Repeat the step more 5 times and collect the flow through into one tube.
(8) Add 5 M NaCl to final concentration of 0.2 M and iso-propanol. Precipitate the cDNA like in stage 1, step 1 and completely dissolve the pellet in 34 μL of water.

3.6 STAGE 5: THE SECOND STRAND CDNA SYNTHESIS

The second strand cDNA synthesis is performed with a thermostable DNA polymerase to ensure efficient synthesis of cDNA regardless of the structure of the cDNA. The 5' of mRNAs, where a substantial part of CAGE tags hit, are notoriously rich in CG and form secondary structures.

(1) Transfer the cDNA to a 0.6 ml tube, and add 7.2 μL of Elongase (Invitrogen) buffer A, 4.8 μL of Elongase buffer B, 6 μL of 2.5 mM dNTPs and 6 μL of 100 ng/μL 2nd strand cDNA primer.
(2) Heat the tube at 65°C for 1 min and pause at 65°C.
(3) Add 2 μL of Elongase (Invitrogen) and quickly mix by vortexing.
(4) Carry out PCR in a thermal cycler as follows: 5 min at 65°C, 30 min at 68 °C, 10 min at 72°C; hold at 4°C until further processing.

(5) Add 2 μL of 0.5 M EDTA, 1 μL of 10% SDS and 2 μL of Proteinase K. Incubate at 45°C for 30 min.

(6) Extract the double stranded cDNA (ds-cDNA) with phenol-chloroform (1:1) and chloroform and back extract using 40 μL of the spin column buffer.

(7) Repeat the spin column step (as for the steps 4–7 in the S-400 subsection of Section 4 above), collecting in total 5 fractions.

(8) Precipitate the ds-cDNA like in stage 1, step 1 and then completely dissolve the pellet in 30 μL of water.

3.7 STAGE 6: PREPARING CAGE TAGS

The class IIs restriction enzymes are used here to cleave DNA at a precise distance outside their recognition sites and produce 2 bases overhangs into the cDNA. These ends can be ligated to linkers having complementary 2 bases NN overhangs. Here, we use MmeI, which cleaves N20/N18 apart from its recognition site in the linker used for the second strand cDNA synthesis. The tags are therefore 20 bp CAGE tags. However, the same protocol can be adapted for the EcoP15I (which has appeared on the market more recently) that cleaves at N25/N27 outside its recognition site, producing longer tags. Protocols with EcoP15I are conceptually the same.

3.7.1 MmeI Digestion

(1) Add 2 μL of S-adenosylmethionine and 10 μL of NEB 10 × buffer and use 3U of MmeI for each 1 μg of cDNA, in a final volume of 100 μL. Incubate for 30 – 60 min at 37°C.

(2) Stop the reaction with 2 μL of 0.5 M EDTA, 2 μL of 10% SDS and 2 μL of Proteinase K. Incubate for 15 min at 45°C.

(3) Extract the cDNA with phenol-chloroform (1:1), chloroform and back extract using 50 μL of 0.1 × TE buffer. To a final volume of 150 μL, add 7.5 μL of 5 M NaCl, 3.5 μL of 1 μg/μL glycogen and 375 μL of ethanol.

(4) Precipitate the DNA like in stage 1, step 1 and completely dissolve the pellet in 2 μL of water.

3.7.2 2nd Linker Ligation

At this point, we need to provide a second linker to PCR amplify the CAGE tag.

(1) Add $4\,\mu$L of the $0.4\,\mu g/\mu$L 2nd linker (Appendix) and $8\,\mu$L of water. Heat it to $65°$C for $2\,$min and quickly place on ice.
(2) Add $2\,\mu$L of $10 \times$ T4 DNA Ligase buffer, $2\,\mu$L of water and $2\,\mu$L of T4 DNA Ligase. Incubate at $16°$C overnight.
(3) Heat the tube to $65°$C for $5\,$min and place on ice. Add $80\,\mu$L of $0.1 \times$ TE buffer.

3.7.3 Purification with Magnetic Beads

At this point, the CAGE tags have to be purified from the "2nd linkers" dimers, which would be overly amplified by PCR. This is achieved by taking advantage of the biotin group at the 5′ end of the second-strand cDNA linker; the "2nd linkers" are not biotinylated and are eliminated. Subsequently, the cDNA is removed from beads by incubating the magnetic beads with free biotin dissolved in guanidine solution; this causes the biotinylated cDNA to be finally released after the bead purification steps.

(1) Prepare Dynabeads M-280 Streptavidine (Dynal) beads by adding $2\,\mu$L of $20\,\mu g/\mu$L tRNA into $200\,\mu$L of beads. Place on ice for $30\,$min and occasionally vortex. Separate beads with magnetic stand and discard the supernatant. Wash the beads 3 times with $200\,\mu$L of $1 \times$ B + W buffer. Resuspend the beads with $100\,\mu$L of $2 \times$ B + W buffer.
(2) Prepare 1.5% free biotin solution in Solution D. In presence of biotin and denaturing condition, the streptavidin-biotin complex can be reverted.
(3) Transfer the washed beads into cDNA solution. After gently mixing, rotate the tube for $15\,$min at room temperature.
(4) Separate beads with a magnetic stand and transfer the supernatant into a new tube. Wash the beads with $200\,\mu$L of $1 \times$ B + W buffer with $1 \times$ BSA (final concentration $0.1\,$mg/ml) twice, $200\,\mu$L of $1 \times$ B + W buffer twice and $200\,\mu$L of $0.1 \times$ TE buffer twice.
(5) Add $50\,\mu$L of 1.5% free biotin in solution D and incubate for $30\,$min at $45°$C to elute DNA from the beads. Separate the beads with a magnetic stand and transfer supernatant into a

collection tube. Repeat this elution step 4 times. Add 50 μL of 0.1 × TE buffer to beads and recover the supernatant. Total volume of eluted DNA solution should be ~250 μL.

(6) Add 12.5 μL of 5 M NaCl, 3.5 μL of 1 μg/μL glycogen and 625 μL of ethanol. Precipitate the DNA like in stage 1, step 1 and dissolve the pellet in 45 μL of 0.1 × TE buffer.

(7) Add 5 μL of 10 × RNaseI buffer and 2 μL of RNase I. Incubate for 15 min at 37°C. Add 1 μL of 0.5 M EDTA, 1 μL of 10% SDS and 1 μL of Proteinase K. Incubate for 30 min at 45°C. Extract with phenol/chloroform, back extract with 50 μL of 0.1 × TE buffer and chloroform. Both the 1st extraction and 2nd extraction measure 50 μL.

3.7.4 Purification by G50 Column

(1) Resuspend the G50 resin in the spin column by vortexing. Loosen the cap and snap off the bottom closure. Place the column in a 2 ml tube and pre-spin it at 735 g for 1 min.

(2) Place the column in a new 1.5 ml tube, slowly apply 50 μL of 1st extract to the top-center of resin. Then add 50 μL from the 2nd extraction.

(3) Spin the column at 735 g for 2 min. The purified cDNA collects at the bottom of the tube. Add 5 μL of 5 M NaCl and 100 μL of isopropanol. Precipitate the DNA like in stage 1, step 1 and then dissolve the pellet in 24 μL of water.

3.8 STAGE: 7 AMPLIFICATION OF CAGE TAGS

It is important to use the lowest PCR cycle number possible to minimize the chance of PCR-induced sequence errors, and most importantly, to minimize representation bias. Additionally, if too many PCR cycles are performed, single-strand DNA may accumulate. These cannot be further cleaved with restriction enzymes, resulting in a loss of material. Theoretically, single-strand complementary tag sequences can hybridize to each other, but this will be directly proportional to their concentration, which leads to the preferential renaturation of highly represented tags, with potential further bias. In summary, it is important to test the number of PCR cycles with aliquots of the reaction before proceeding with the bulk PCR.

3.8.1 1st PCR Amplification

(1) Set up at least 3 reactions to decide the optimal number of PCR cycles (for instance, 18, 20, 22 cycles; this may vary depending on the experience of the operator).

10 × PCR buffer	5.0 μL
DMSO	3.0 μL
2.5 mM dNTPs	12.0 μL
Primer1 (350 ng/μL)	0.5 μL
Primer2 (350 ng/μL)	0.5 μL
Template	1.6 μL
Water	26.6 μL
DNA Polymerase (2.5 U/μL)	0.8 μL
Total volume	**50.0 μL**

PCR conditions:
Step 1: 94°C, 1 min
Step 2: 94°C, 30 sec
Step 3: 55°C, 20 sec
Step 4: 70°C, 20 sec
Go to Step 2, 18/20/22 cycles
Step 5: 70°C, 5 min
Step 6: 4°C, hold

(2) Run the products on 12% PAGE gel to identify the optimal PCR cycle number to produce a PCR band that does not reach saturation.

(3) Set up new PCR reaction with 10 tubes, and perform the PCR as in step 1 above. The remained sample should be stored at −20°C in case you have to utilize the sample again.

(4) Collect the samples (ten tubes are pooled into a 1.5 ml tube); the total volume should be 500 μL.

(5) Add 10 μL of 0.5 M EDTA, 10 μL of 10% SDS and 10 μL of Proteinase K. Incubate at 45°C for 30 min. Extract with phenol/chloroform, back extract with 50 μL of 0.1 × TE buffer and chloroform. Add 5 M NaCl and isopropanol. Precipitate like in stage 1, step 1 and dissolve the pellet in 24 μL of 0.1 × TE.

3.8.2 PAGE Purification

(1) Mix the purified DNA with the loading buffer and apply to 12% 1 × TA polyacrylamide gel.

(2) Cut the band to desired size band (65 bp with the described primer design) and put the gel in a 1.5 ml tube. Completely crush the gel and add 150 μL of elution buffer. Rotate at room temperature overnight.

(3) Centrifuge at 15,000 rpm and transfer the supernatant into empty MicroSpin Columns. Centrifuge at 3000 rpm and collect the flow through in a new 1.5 ml tube.

(4) Add 150 μL of elution buffer to the gel and rotate the tube at room temperature for 30 min. Repeat this elution step 4 times.

(5) Extract with phenol/chloroform and chloroform to eliminate the remaining gel. Perform the back extract with 50 μL of 0.1 × TE buffer. Precipitate the DNA with ethanol and dissolve the pellet in 30.5 μL of 0.1 × TE. Check the concentration with 0.5 μL of sample solution, for example, by using picogreen.

3.8.3 2nd PCR Amplification

(1) This step is important in obtaining more DNA; this is mandatory for the 454 and for other methods where concatenation steps are essential. As commented above for the first PCR (beginning of Stage 7), It is very important to avoid overcycling. Usually 6 to 10 cycles are enough at this stage.

Fortunately CAGE is being modified for other sequencing instruments (Illumina/Solexa and SOLiD) and we believe that the second PCR step described below will become unnecessary to achieve sufficient DNA for sequencing.

(1) Set up 3 reactions to decide the PCR cycles number (for instance, 6, 8, 10 cycles).

10 × PCR buffer	10.0 μL
DMSO	6.0 μL
2.5 mM dNTPs	12.0 μL
Primer 1 (1 μg/μL)	0.75 μL
Primer 2 (1 μg/μL)	0.75 μL
template	0.2-1 ng

DNA Polymerase (2.5 U/μL)   0.8 μL
Final volume   **100.0 μL**

PCR conditions are as follows:
Step 1: 94°C, 1 min
Step 2: 94°C, 30 sec
Step 3: 55°C, 20 sec
Step 4: 70°C, 20 sec
Go to Step 2, 6/8/10 cycles
Step 5: 70°C, 5 min
Step 6: hold at 4°C

(2) Select the PCR condition by running a 12% PAGE gel.
(3) And set up the reaction with 50–100 additional tubes.
(4) Collect the 5 of the samples above into each one 1.5 ml tube. The volume per each tube is 500 μL.
(5) Add 10 μL of 0.5 M EDTA, 10 μL of 10% SDS and 10 μL of Proteinase K. Incubate at 45 °C for 30 min. Extract with phenol/chloroform, back extract with 50 μL of LoTE buffer and chloroform. Add 5 M NaCl and isopropanol. Precipitate like in stage 1, step 1 and dissolve the pellet in 25 μL of 0.1 × TE. Pool the content of the 2 tubes into 1 tube. The total volume of 1 tube should be ~50 μL.

3.8.4 *Purification with QIAGEN MinElute Column*

This step is necessary to eliminate the remaining primer that may interfere with subsequent concentration measurements.

(1) Perform the purification as suggested in the protocol attached to the Qiagen MinElute kit, following by DNA pooling and concentration by ethanol purification.
(2) Quantify the purified DNA using picogreen on a small aliquot of the purified PCR bands.

This is a convenient stopping point; samples should be stored at −20°C.

3.9 STAGE 8: RESTRICTION

When concatenation if needed (like in the case of Sanger sequencing or 454 Life Science sequencing), the tips of the PCR products should be removed and only the central part of the PCR products, containing the CAGE tags (and optionally the tissue barcode tags), should be concatenated. This involves restriction digestion followed by removal of the tips of the linkers with Streptavidin beads and then electrophoresis.

3.9.1 Restriction with XmaJI

(1) Divide the sample into multiple tubes (use 2 μg DNA for each digestion reaction).
(2) Add 20 μL of 10 $\times$ XmaJI buffer, 18 μL of XmaJI (10U/ul) and ddH$_2$O up to 200 μL. Incubate at 37°C for 2 hours. Then take an aliquot and run on 12% PAGE gel to check that restriction is complete. If the reaction is not complete, incubate for additional time (1 – 2 hours).
(3) Add 4 μL of 0.5 M EDTA, 10% SDS and Proteinase K. Incubate at 37°C for 30 min.

3.9.2 Removal of the Linkers Tips

(1) Prepare the Streptavidin beads M-280. Put 200 μL of bead solution into a 1.5 ml tube. Wash the beads with 200 μL of 1 $\times$ B + W buffer 3 times. Resuspend the beads with 100 μL of 2 $\times$ B + W buffer.
(2) Add the washed beads to sample. Gently rotate for 15 min at room temperature.
(3) Separate the beads with a magnetic stand for 2-3 min and collect the supernatant in a 1.5 ml tube.
(4) Add 75 μL of 1 $\times$ B + W buffer and collect the supernatant in the tube.
(5) Extract with phenol/chloroform to eliminate the contaminated beads, back extract with 50 μL of LoTE and chloroform. Precipitate and dissolve completely the pellet in 45 μL of TE.

3.9.3 PAGE Purification

(1) Mix purified DNA with loading buffer, apply to 12% TA polyacrylamide gel and separate.
(2) Cut the band (32 bp or 37 bp depending on the design) and transfer the gel in a 1.5 ml tube. Completely crush the gel and add 150 μL of elution buffer. Rotate at room temperature overnight.

(3) Centrifuge at 15,000 rpm and transfer the supernatant into MicroSpin Columns. Centrifuge at 3000 rpm and collect the flow through in a new 1.5 ml tube.

(4) Add 150 μL of elution buffer to the gel and rotate the tube at room temperature for 30 min. Repeat the elution step 4 times.

(5) Extract with phenol/chloroform to eliminate remaining gel, back extract with 50 μL of 0.1 × TE buffer and chloroform. Total volume should be around 650 μL.

(6) Divide the sample into 2 tubes and add 17 μL of 5 M NaCl and 325 μL of ethanol. Precipitate like in step 1 and dissolve the pellet in 6.5 μL of 0.1 × TE. Check the concentration with 0.5 μL of sample solution by picogreen (Invitrogen).

3.10 STAGE 9: CONCATENATION

Sanger sequencing has been typically used to read CAGE libraries,[1] but due to the cost, this is not used any more. In our experience, at least 500 ng of tags is necessary for the concatenation in GS20/GSFLX systems (454 Life Sciences). The concatenation takes place by addition during the ligation steps with the "454" linkers; these can ligate only at one of their sides and act as "ligase terminators". These products are then separated to sequence only large concatamers.

(1) Add 1/20 amount of tags of 454 adaptor A/B, 1.0 μL of 10 × T4 DNA ligase buffer, 1.0 μL of T4 DNA ligase and water up to 10 μL to 6 μL of CAGE tags.

(2) Incubate overnight at 16 °C.

(3) Add 1 μL of 0.5 M EDTA, 1 μL of 10% SDS and 1 μL of proteinase K.

(4) Incubate at 45°C for 30 min.

(5) Mix concatemer with loading buffer, apply to 2% agarose gel.

(6) Cut out gel of 300 bp over, and purify the DNA in the agarose gel with the QIAGEN gel purification kit. Next, the concatemer ligated with 454 adaptors should be denatured into ssDNA with GS20/GSFLX system (454 Life Sciences) kit followed by emulsion PCR according to the instructions of the manufacturer.

Future challenges include the simplification of various steps of these protocols. In particular we hope to avoid multiple electrophoresis and decrease the number of PCR amplifications. This will be achieved with the Illumina, Solid by decreasing the cycle number. Additionally this may be achieved in the third generation sequencers (single molecule sequencers).

3.10.1 Appendix

Below is the information regarding reagents and solutions used in this protocol.

Solutions and kits
Sorbitol/Trehalose mix: 2:1 (vol/vol) of 4.9 M sorbitol and saturated solution (80%) of D-(+)-trehalose (>99.5% by HPLC; Fluka)

CTAB/urea solution: 1% cetyltrimethylammonium bromide (CTAB; cationic detergent), 4M urea, 50 mM Tris-HCl (pH 7.0), 1 mM EDTA

Biotin (long arm) Hydrazide (Vector Laboratories)

S400 MicroSpin Column (GE Healthcare)

spin column buffer: 10 mM Tris-HCl (pH 8.0), 0.1 mM EDTA, 0.1% SDS, 100 mM NaCl

S400 HR matrix (GE Healthcare)

open tip spin column (Assist)

1 × B + W buffer: 5 mM Tris-HCl (pH 7.5), 0.5 mM EDTA, 1M NaCl

2 × B + W buffer: 10 mM Tris-HCl (pH 7.5), 1 mM EDTA, 2M NaCl

Solution D: 4M guanidium thiocyanate, 0.5% n-lauryl-sarcosine, 25 mM sodium citrate (pH 7.0)

G50 spin column (GE Healthcare)

MicroSpin Column (GE Healthcare)

PAGE elution buffer: 2.5 mM Tris-HCl (pH 7.5), 1.25 M ammonium acetate, 0.17 mM EDTA (pH 7.5)

picogreen (Invitrogen)

5′ linker
A specific linker, containing a recognition site for XmaJI and a Tissue Id tag (5bp) and the class II restriction enzyme MmeI ("upper oligonucleotide GN5" sequence: biotin-agagagagagacctcgagtaactataacggtcctaaggtagcgacctaggXXXXXt-ccga cGNNNNN, and "upper oligonucleotide N6" sequence: biotin- agagagagagacctcgagtaactataacggtcctaagg tagcgacctagg

XXXXXtccgacNNNNNN,:XXXXX: Tissue barcode ID tag) are mixed in a ratio of 4:1, and then this mixture in turn was mixed at 1:1 to the "lower oligonucleotide" sequence: Pi-gtcggaXXXXXcctaggtcgctaccttaggaccgttatagttactcgaggtctc tctct-NH2). Add NaCl to a final concentration of 0.1 M, then heat the mixture to 95°C and cool down gradually to anneal the linker's up strand with the down strand.

3' linker
A specific linker, containing a recognition site for XmaJI (up: Pi-cctaggtcaggactcttctatagtgtcacctaaagacacacacac-NH2, and down:gtgtgtgtgtctttaggtgacactatagaagagtcctgacctaggNN) is mixed in a ratio of 1:1. Add NaCl to a final concentration of 0.1 M, heat to anneal them, then gradually cool down.

Oligonucleotides
Random
primer: 5'-NNNNNNNNNNNNNNNNNNNNN-3'Oligo-dT primer: 5'-TTTTTTTTTTTTTTVN-3' return 2nd primer: 5'-biotin-AGAGAGAGACCTCGAGTAACTATAAC3'
Primer 1: 5'-biotin-CTATAGAAGAGTCCTGACCTAGG-3'
Primer 2: 5'-biotin- CGGTCCTAAGGTAGCGACCTAG-3'

Note that the composition of linkers can be changed depending on the how the protocol is revised for different sequencing platforms.

References

[1] R. Kodzius *et al.* CAGE: cap analysis of gene expression. *Nat. Methods* **3**, 211–222 (2006).

[2] M. Margulies *et al.* Genome sequencing in microfabricated high-density picolitre reactors. *Nature* **437**, 376–380 (2005).

[3] P. Carninci *et al.* High-efficiency full-length cDNA cloning by biotinylated CAP trapper. *Genomics* **37**, 327–336 (1996).

[4] P. Carninci *et al.* Genome-wide analysis of mammalian promoter architecture and evolution. *Nat. Genet.* (2006).

[5] N. Maeda *et al.* Development of a DNA barcode tagging method for monitoring dynamic changes in gene expression by using an ultra high-throughput sequencer. *Biotechniques* **45**, 95–97 (2008).

Chapter Four

Transcriptome and Genome Characterization Using Massively Parallel Paired End Tag (PET) Sequencing Analysis

Chia-Lin Wei and Yijun Ruan*

Genome Institute of Singapore, Singapore
*Email: *ruanyi@gis.a-star.edu.sg*

The next generation of DNA sequencing technology has had a widespread impact on many aspects of biological research. Though the ultra high throughput and large volume production of DNA sequence data by the new sequencing platforms is powerful, the short read length nature is a significant limiting factor to its applications for many genomic analyses. An immediate and widely recognized solution to overcome the problems associated with short tag-based sequencing platforms is to adapt the paired end ditag (PET) sequencing strategy. It has been demonstrated that multiplex sequencing of paired end ditags (MS-PET) is not only a very efficient and effective strategy to extend the linear coverage of tag-based sequencing methods, but also an unique approach for the comprehensive discovery of unconventional fusion transcripts and rearrangement events from genomic DNA structures, and for whole genome mapping of *in vivo* DNA protein binding and long range chromatin interactions. Collectively, coupled with the ultra high throughput short tag sequencing platforms, the PET technology can have broad applications for a wide range of genomic analyses.

Cap Analysis Gene Expression (CAGE): The Science of Decoding Gene Transcription edited by P Carninci
Copyright © 2010 by Pan Stanford Publishing Pte Ltd
www.panstanford.com
978-981-4241-34-2

4.1 INTRODUCTION

With the complete human genome sequences in hands[16,40] we are now in the starting points of identifying all functional genetic elements encoded in the human genome, and begin to elucidate the complex regulatory networks that coordinate functions of all genetic and epigenetic elements[2] throughout development and disease progression. Furthermore, with the availability of the reference and a number of individual human genome sequences, we are able to analyze the genetic variations in our entire genetic makeup in order to understand their impacts in our phenotype variations and determining our susceptibility to diseases and drug treatments. All these promises, in fact, require huge sequencing capability and advanced sequencing strategy to rapidly decode millions or billions of DNA in high efficiency.

The recently developed next generation DNA sequencing technologies, represented by 454/Roche, Solexa/Illumina and Agentcourt/ABI's SOLiD platforms,[20,29] have triggered a new wave of revolution in many aspects of genomics and biological research. A number of other DNA sequencing platforms are also on their way to the market place, including Helicos' and Pacific Bioscience's single DNA molecule sequencing platforms. Though these new sequencing systems are, in fact, quite different from one another and at various stages of maturity, the major advance of these new sequencing technologies is their ability to massively generate large volume of DNA sequence data from hundreds of mega-bases up to several billion bases per machine run without the requirements of tedious of DNA cloning and laborious preparation of sequencing templates. However, the obvious weak point shared by all of the current new sequencing technologies is the short read-length (100–250 bp by 454's GSFLX; 25-35bp by Solexa, SOLiD or Heliscope). Despite the limitation of short read length, highly multiplex nature of these new sequencing methods already generated tremendous impact.

An immediately and widely recognized solution to overcome the problems associated with short tag-based sequencing platforms is to adapt the paired end ditag (PET) sequencing strategy that we had originally developed for cloning-based transcriptome analysis[25] and whole genome mapping of transcription factor binding sites.[35] We have demonstrated that the PET scheme can be easily adapted to the tag-based multiplex sequencing platform,[24] in which PET can overcome for the inherent limitations of short reads by providing paired-end information from long contiguous DNA fragments. The multiplex sequencing of paired-end ditags (MS-PET) not only extends the linear DNA sequence coverage, but more importantly, enables the inferring of the relationship between the two ends of DNA fragments in defined distance and content. This unique feature has enabled us to identify unconventional fusion transcripts[27] and genome structural variations.[14] Therefore, the MS-PET sequencing strategy not only can offer unique advantage to improve the efficiency and accuracy of short tag-based sequencing methods, but also expand their applications and information outputs. Collectively, the PET technology shown here has tremendous and immediate value in providing unique and comprehensive solutions in this fast growing research era for whole genome characterization of transcription and epigenetic regulatory networks and whole genome scan for chromosomal structural variations.

Here, we describe the concept of the MS-PET approach and its applications in transcriptome analysis, mapping the DNA-protein interactions, and identifying genomic structural variations. The potential of applying PET analysis on re-sequencing the human genomes and its implications in personalized genome medicine are also discussed.

4.2 THE DEVELOPMENT OF PAIR END DITAG (PET) ANALYSIS

Our interest started with the use of full length cDNA cloning[5] coupled with SAGE/MPSS[4,28] sequencing approaches to characterize mammalian transcriptomes. It is well known that the full length cDNA approach is too expensive and labor intensive, while the SAGE approach, although efficient in counting cDNA tags, lacks the specificity in characterizing transcripts, especially where the transcript starts (5′ end) and terminates (3′ end). To improve our capability for comprehensive transcriptome analysis, we first developed 5′LongSAGE and 3′LongSAGE protocols to map transcription start sites (TSS) and polyadenylation sites (PAS) of gene[35] Expanding from such capability, we then devised the paired-end ditagging (PET) strategy, in which the 5′ and 3′tags derived from a full length cDNA fragment are covalently linked as a single molecule for sequencing analysis.

The principal concept of the PET methodology is to extract the paired end signatures from each of the target DNA fragments, and map the paired tag sequences to the reference genome for accurate demarcation of the boundaries of the tested DNA fragments on the genome landscape. Starting from any kind of DNA (cDNA or genomic DNA), in this process, specific adapter sequences are ligated to the DNA fragments,followed by Type IIs restriction enzymes' digestions to release the paired end ditags (PETs), with tag from each end of 20 bp in length. The connectivity of the two paired tags is achieved through the use of cloning vector that embrace the insert DNA fragment[25] Later, we simplified this step by circularization of the adapted DNA fragments (the *in vitro* cloning free method) (Fig. 4.1). The PET structure containing two paired short tags can then be either concatenated into longer stretch of 10–20 PET sequence units for efficient high throughput

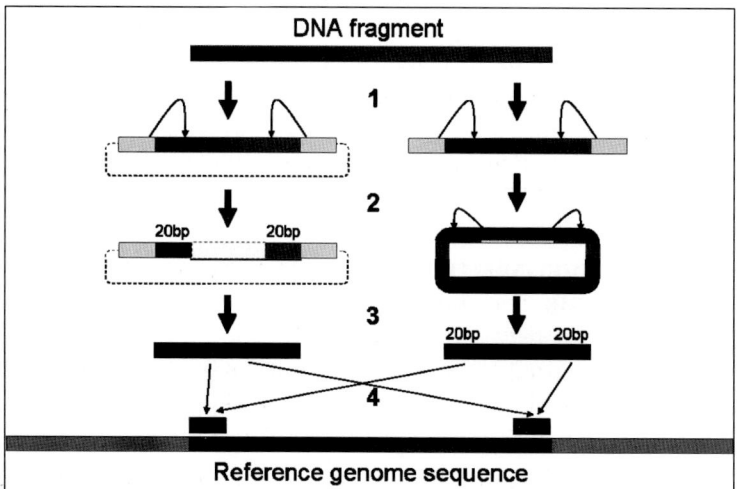

Figure 4.1. The schematic view of paired end ditag analysis Left: cloning-based PET method. Adaptors ligated DNA fragments are cloned into the vector before they are subjected to type IIs restriction enzyme digestion. Right: the cloning-free PET approach. Adaptor ligated DNA fragments are self-circularized followed by type IIs restriction enzyme digestions. The resulted PETs are sequenced and map to genome to determine the identity of DNA fragments of interest.

sequencing by traditional capillary method or directly multiplex sequenced by the next generation sequencing platforms. The PET sequences are then characterized by mapping to the reference genome sequences. The paired 5′ and 3′ signatures were considered mapped if they were located on the same chromosome, same strand (+ or −), 5′ tag location followed by the 3′ tag location and within the expected genomic distance of each other. As the results of such mapping criteria, vast majority of these PETs can be uniquely located to the reference genome and accurately define the identity of the DNA fragments analyzed.

Based on the PET concept, the Gene Identification Signature (GIS) analysis using paired end ditags (GIS-PET) was developed to characterize full length cDNA fragments and transcriptome. The full length cDNA fragments are first cloned into vector to generate full length cDNA library, then the cDNA inserts are digested by *Mme*I restriction enzyme specifically incorporated into the cloning junctions, and the two tags remained on the cloning vector are jointed through circularization ligation reaction. The resulting single PET constructs are concatenated into longer DNA fragments for high throughput sequencing by traditional capillary method. When mapped to the genomic sequences, the PET sequences can precisely demarcate the boundaries of full-length transcripts on genome landscape.[25] Immediately afterward, the PET strategy was adopted to analyze the ChIP fragements as the ChIP-PET analysis (chromatin immuno-precipitation coupled with paired-end ditagging) for highly accurate, robust and unbiased genome-wide identification of transcription factor binding sites.[35] Through the availability of 454 sequencing technology, this new sequencing method was quickly adapted by the PET approach as the multiplex sequencing for paired-end ditag (MS-PET) analysis.[25] Such strategy achieved additional 100-fold efficiency compared to the use of conventional sequencing method for PET experiments.

4.3 GIS-PET FOR TRANSCRIPTOME ANALYSIS

Traditionally, transcripts and transcriptomes are studied by DNA microarrays and cDNA sequencing. The advancement of DNA microarray fabrication that covers the entire genome is an attractive approach to study transcriptome. The genome-wide tiling arrays[30] provided a massive parallel approach for characterization of all expressed exons and new promise for highly comprehensive transcriptome analysis. However, the tiling array data provides no connectivity for structural information or strand specificity for each transcript to be characterized; i.e. it is not straightforward to define the start and termination positions of individual transcript units and the connectivity of each exons. Furthermore, the tiling array approach suffers from cross hybridization noise when it is used to detect transcripts expressed in highly homologous genomic regions. In contrast, the sequencing-based strategies, such as full-length cDNA sequencing[32] and short tag sequencing of SAGE[28,33] and MPSS[4] had contributed immense volume of transcriptome data, but were limited by either huge operational cost, inefficiency (full-length cDNA sequencing), or by insufficient information (SAGE and MPSS tags) (Fig. 4.2).

In GIS-PET analysis, the transcriptome of a biological sample is first constructed as a full-length cDNA library that captures all intact transcripts. This library is then subjected for PET analysis. PETs from the two ends of each expressed full length transcript (18 bp from 5′ end and 18 bp from 3′ end) are extracted and subjected to multiplex sequencing (Fig. 4.2). Millions of transcripts represented by PETs can be analyzed at high efficiency and the PET sequences are precisely mapped to the genome for the identification of expressed genes and quantification of expression levels. We have demonstrated that PET sequences can accurately demarcate the 5′ end and the 3′ end of the individual full-length transcripts of different alternative forms expressed, and the copy numbers of PETs mapping to specific loci provide digital counts of gene expression level. In addition to analyze expressed transcriptomes and accurately demarcate gene transcription boundaries, GIS-PET can also infer proximal promoter sites and enable the discovery of novel genes and alternative transcript variants with unprecedented efficiency. Furthermore, this pair-end property enables the identification of unconventional transcripts such as those formed by intergenic, bicistronic linkage or trans-splicing events, which can not be uncovered using the array-based approach. These efficient features make GIS-PET as the ideal approach for gene identification and differential quantification as demonstrated in the FANTOM project for comprehensive mouse

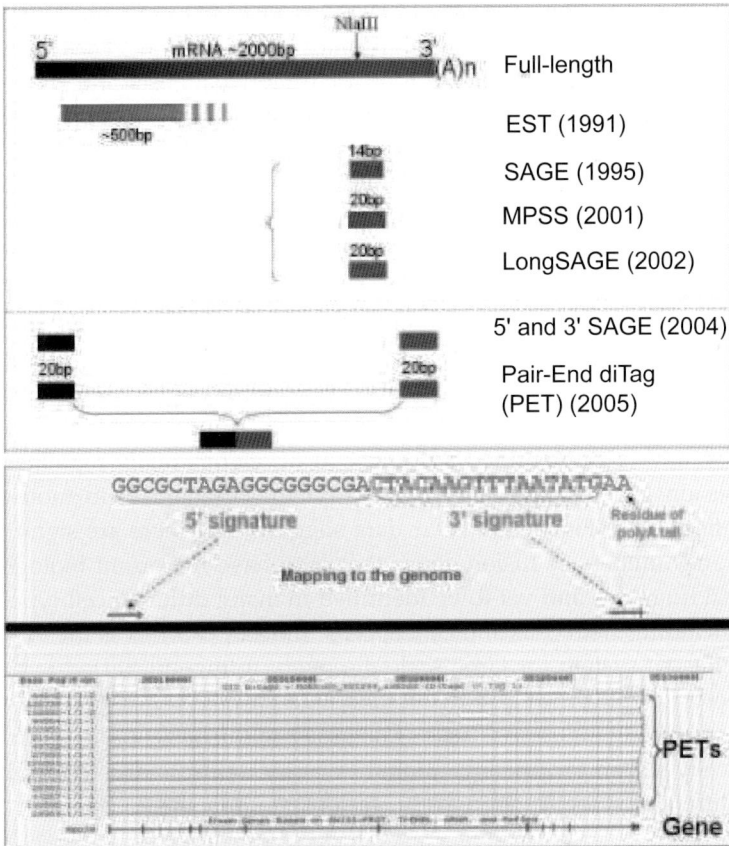

Figure 4.2. GIS-PET for transcriptome analysis Top: Diagram describes various sequencing strategies used for transcript identification and profiling, when they became available and the information content these methods provided. Pair-end ditags were developed by connecting the 5′ and 3′ tags from each end of full length cDNA fragments. Bottom: Example of a PET structure, 5′ and 3′ tags mapped to reference genome to define the TSS and PAS of expressed transcripts and the numbers of PETs representing transcript abundance.

transcriptome analysis[6] and in the ENCODE project to characterize the 1% human genome for functional DNA elements.[3]

The most unique feature of GIS-PET is its ability to delineate the relationship between the two ends of individual cDNA molecule. Therefore, GIS-PET is unique and efficient for discovery of fusion genes resulted from genome rearrangements or trans-splicing (Fig. 4.3). through PET mapping on different orientations or different chromosomes. Fusion genes have been shown as a valuable tool for tumor diagnosis and therapeutic stratifications and there is a variety of mechanisms used to generate fusion transcripts with novel functions.[23] *BCR-ABL* translocation, the well known example, was used successfully in the discovery of new diagnostic marker and the development of Gleevec® in CML.[21] We have applied the GIS-PET to characterize the transcriptomes from two of the well studied cancer cell lines, breast cancer MCF7 and colon cancer HCT116, and successfully identified more than 70 potential fusion genes.[27] One of them, *BCAS4/BCAS3*, has been verified through independent cytogenetic and genomic DNA sequencing approaches (Fig. 4.3). We have also identified a number of developmentally regulated fusion genes derived from trans-splicing mechanism in mouse embryonic stem cells.[25] Our data suggests that there are significant numbers of fusion genes that may have important biological functions and yet to be identified in many different cell types including stem cells and cancer cells. Indeed, GIS-PET is the only efficient system for large scale discovery in this uncharted territory. Coupled the GIS-PET method with the next generation DNA sequencing platforms, large scale program can be set up to specifically screen for unconventional fusion transcripts derived from different biological systems by various mechanisms.

4.4 CHIP-PET FOR WHOLE GENOME MAPPING OF TRANSCRIPTION FACTOR BINDING SITES AND EPIGENETIC MODIFICATIONS

It is known that gene transcription in eukaryotic cells is regulated by specific transcription factors with specific DNA recognition properties through direct or indirect binding to the regulatory DNA elements in a spatial and temporal specific manner. Thus, the identification of functional elements such as transcription-factor binding sites (TFBS) on a whole-genome level is the next challenge for genome sciences and gene-regulation studies. Increasing evidence also suggests that many TFs function in a cooperative manner to form complex of transcription regulatory circuits. However, how to study such complex regulatory networks has been a big challenge of biology. These

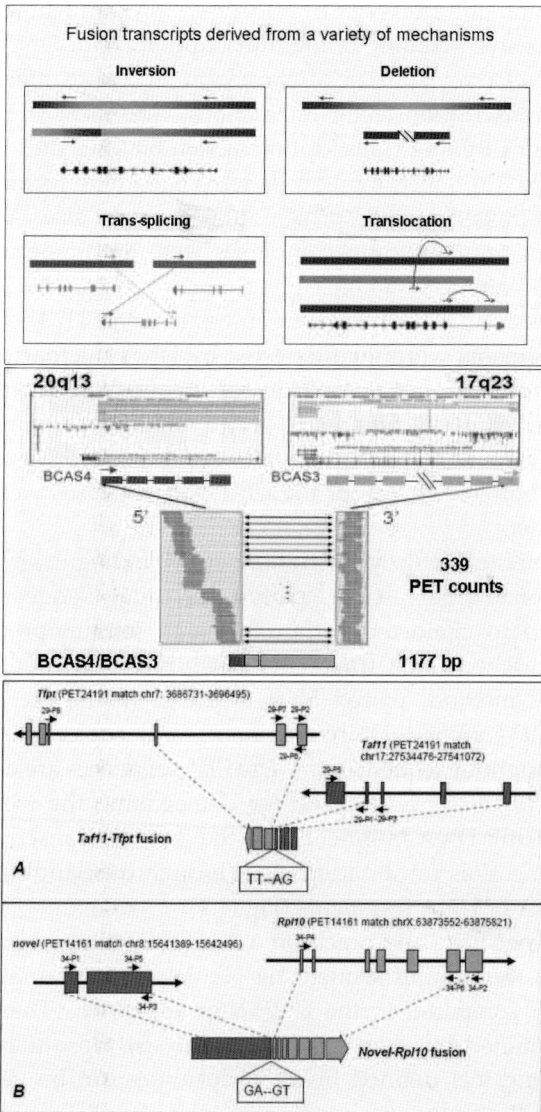

Figure 4.3. Unconventional fusion transcripts identified by GIS-PET analysis Top: Types of fusion transcripts can be identified by GIS-PET analysis and the resulted PETs mapping patterns on the reference genome.Middle: *BCAS4/BCAS3* fusion genes uncovered by GIS-PET analysis from MCF7 cells. Multiple PET clusters (339 PETs) with their 5′ tags mapped to chr20q13, the starting region of *BCAS4* gene whereas the 3′ tags mapped to chr17q23, the 3′ terminal region of *BCAS3* gene. The resulted fusion transcript is 1177 bp. Bottom: Two different fusion genes derived from trans-splicing events in mouse embryonic stem cells.

fundamental questions can be addressed through whole genome comprehensive profiling of the transcription factor DNA interactions and characterization of chromatin structures mediated by specific transcription factor interactions. The most widely used method for profiling transcription faction binding sites had been ChIP-chip,[36] in which living cells were fixed with formaldehyde that cross-links the DNA/protein interactions *in vivo*. After fragmentation, the chromatin complexes were immunoprecipitated by specific antibody against given protein factor and therefore enrich the target DNA fragments associated with TF. The enriched DNA fragments were then detected by DNA microarray.[15] Due to the large size and complexity of mammalian genomes, the DNA microarrays constructed were often containing partial genomic content or only promoter regions of well characterized genes. Therefore, many of the ChIP-chip analyses were in fact incomplete.

In order to profile transcription factor binding in an unbiased and genome-wide fashion, paired end ditag sequencing was developed to characterize the chromatin immunoprecipitation (ChIP) enriched DNA fragments as ChIP-PET method.[35] In the ChIP-PET method, paired 5' & 3' ditags (PETs) are extracted from ChIP-enriched DNA fragments and subjected for ultra high throughput sequencing.[24] The PET sequences are accurately mapped to reference genome for demarcating the locations of inferred ChIP DNA fragments and the genuine transcription factor binding sites can be identified through overlapping of PET-inferred ChIP DNA fragments (Fig. 4.4).

The specificity and precision of ChIP-PET method has proved that high throughput sequencing is a superior readout to identify TFBS compared to the ChIP-chip approach. Through cut-off determined by simple statistic analysis, more than 99% of the binding loci defined by PET clustering can be verified by ChIP-qPCR validation experiments, and the PET-defined binding regions can be narrowed down to less than 10 bp,[35] which allows the *de novo* deriving of the consensus binding sequence motif. Using ChIP-PET, a number of important transcription factors, including p53,[35] Oct4 and Nanog[19] cMyc;[37] ERá[18] and NF-?B[17] have been successfully mapped to the mouse and human genomes as well as the epigenomic profiles of histone modifications in human embryonic stem cells.[39]

Encouraged by the ChIP-PET results, the sequencing-based approach for TFBS analysis has been elevated to a new level in

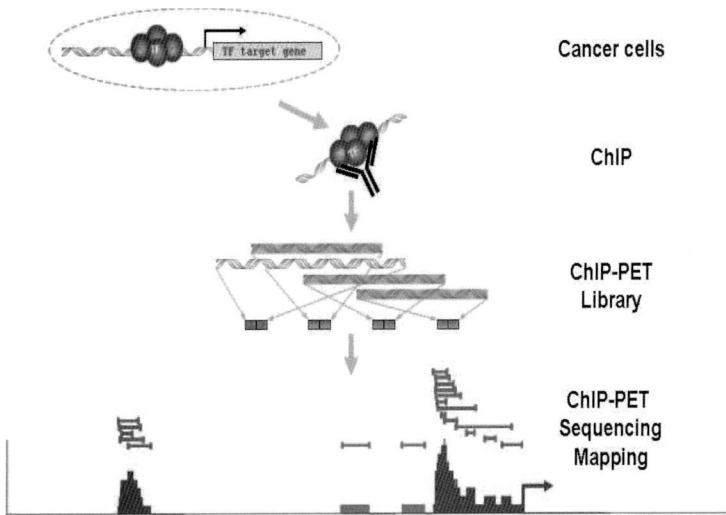

Figure 4.4. ChIP-PET to map transcription factor binding sites by chromatin immunoprecipitation. PETs from these DNA can be extracted, sequenced and mapped to the genome. The genomic loci shared by multiple PET mapping regions can be inferred as TFBS.

2007 through the available of Solexa sequencing platform. The advantages contributed by Solexa sequencing in ChIP analysis are its massive output of short tag sequencing reads (>40 million tag reads of 36 bp per machine run) and its requirement for small amount of template DNA (ng level). These advantages are immediately appreciated and recognized as a new method ChIP-Seq, in which ChIP DNA fragments are directly end sequenced to generate millions of short tags to identify TFBS. The application of ChIP-Seq has generated exciting results in mapping histone modifications[1,2] and TFBS[13] These reports have reinforced that sequencing is a simple and robust approach for readout ChIP DNA and globally profiling of the transcription factor binding sites with high specificity and accuracy. In summary, sequencing-based approach is the superior approach for whole genome protein/DNA interaction analysis, and the accumulation of such knowledge will be useful to systematically construct the transcriptional networks and regulatory circuitries. Such global approach should provide invaluable information to decipher the gene regulation programs.

4.5 CHIA-PET FOR WHOLE GENOME IDENTIFICATION OF LONG RANGE INTERACTIONS

As discussed above, significant progress has been made in identifying genes and regulatory elements that modulate transcription and replication of the genome. A growing body of data generated by various whole genome TFBS studies[3,7,19,35,37] has started to show that a large portion of the putative regulatory elements are localized far away from genes coding regions; this phenomenon is difficult to be explained by the simple linear relationship of locations between such elements along the genome.

It has been hypothesized for years that through tertiary conformation of chromatin, DNA elements can function at considerable genomic distance from the genes they regulate.[12] Studies of *in vivo* chromatin interactions in specific cases have demonstrated that enhancers and locus control regions (LCRs) spanned long-range (up to megabase scale) can be found in close spatial proximity to their target genes.[8,26] Emerging data from recent studies also raises the possibility that chromosomes can interact with each other to regulate transcription in *trans*[31] and suggests that higher-level interactions of DNA elements in nuclear 3-dimensional space are important and abundant mechanisms for the regulation of genome functions.

Most of the recent advances in the understanding of chromatin interactions has relied on the Chromosomal Conformation Capture (3C) method,[9] which has proven to be very useful in determining the identity of DNA sequences from remote genomic locations that lie in close physical proximity in formaldehyde-cross-linked chromatin. However, this method is limited to the detection of only specific interactions for which we have prior knowledge or perception of their existence. To overcome this limitation, a number of groups have developed chromosome conformation capture on chip or by cloning, 4C[39] and 5C[10] methods to expand the scope for the detection of such chromatin interactions (Fig. 4.5). Based on 3C circularized ligation products, the 4C approach uses PCR to prime on known targets and subsequently extends into DNA fragments of unknown regions. The 4C-amplified products can then be characterized by either microarrays or sequencing analysis. Hence, the 4C method has the potential to detect many chromatin interactions *de novo* from

a known site (Fig. 4.5). Similarly, starting from the 3C ligation products, the 5C method uses multiplex oligonucleotide
primers pre-designed around many locations of the restriction
sites used for 3C analysis over a large genomic region, and uses
ligation-mediated amplification (LMA) to amplify the 3C products. The 5C amplicons can then be analyzed by specifically-
designed microarrays or DNA sequencing. The 5C primers are
based on restriction sites instead of known targets; therefore it has
the potential to detect chromatin interactions *de novo* (Fig. 4.5).
However, the current version of 5C is constrained by the lim-
its of reliable LMA, and thus only for partial genome. There is
no doubt that the 4C and 5C methods will trigger a new wave
of interest into the long-range control of transcription regulation.
However, these methods essentially rely on the same concept of
PCR selection for targeted interactions and are therefore unable to
achieve unbiased whole genome analysis for the *de novo* discovery of chromatin interactions. To move this field further forward
into 3-dimensional space, it is desirable to develop an unbiased,
whole-genome approach for the *de novo* discovery of chromatin
interactions that is independent of any prior knowledge or information about any sequence feature, especially those involved in
transcriptional regulation.

We therefore devised the Chromatin Interaction Analysis
using Paired End diTag (ChIA-PET), (Fig. 4.5). a genome-wide,
high-throughput, unbiased, and *de novo* approach for detecting
long-range chromatin interactions in 3D nuclear space. This
method combines the "proximity ligation" principle used in the
3C approach[9] with the power of Paired End diTags (PETs)[35] and
next generation sequencing technologies.[20,29] Briefly, a specially
designed DNA oligonucleotide adaptor is introduced to link different DNA fragments that are non-linearly related in the genome
but brought together in close spatial proximity by protein factors
in vivo. The proximity ligation products are then digested by designated type IIS restriction enzyme (MmeI or EcoP15I) to release
paired end ditag structures from each of the ligated DNA fragments. The ditags are subjected to high throughput sequencing,
and the PET sequences are mapped to the reference genome to
reveal the relationship between the paired DNA fragments and
to identify long range interactions among DNA elements. Therefore, by design, the ChIA-PET approach can provide two types
of information, the self-ligation PET that define the transcription

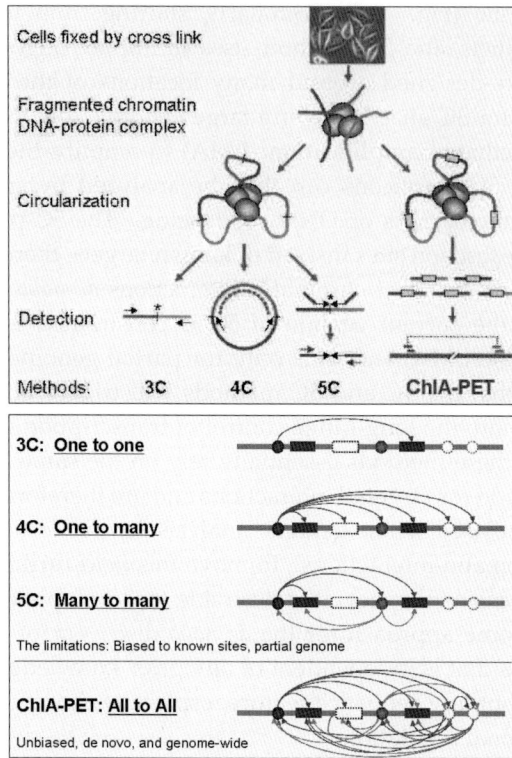

Figure 4.5. Illustration of different approaches used to characterize long range chromatin interactions **Left.** Tethered DNA bound by specific protien factors are ligated. Chromatin interactions can only be detected by PCR using specific PCR primers against the known regions (3C), sequencing analysis of DNA regions from primers extended from specific regions (4C) or ligation mediated PCR from specific designed PCR primers (5C). Right. In ChIA-PET analysis, adaptors containing type IIs RE are ligated at each end of tethered DNA. PETs from interacting chromatin DNAs are generated by RE digestion (MmeI or EcoP15I) and sequenced to determine the interactions at genome wide scale. Below: the scale, resolution and efficiency of ChIA-PET analysis compared with other similar approaches.

factor binding sites and the inter-ligation PETs that reveal the interactions between the binding sites, while the 3C method is only for detection of one-to-one interaction and 4C is one-to-many (Fig. 4.5).

We have begun to demonstrate this approach in the model yeast and in the human genome. Through ChIA-PET experiments, we characterized the estrogen receptor (ERá) mediated interactions in human breast adenocarcinoma cells (Fig. 4.6). In this work, we provided the first global view of long range chromatin interactions mediated by transcription factors in the human genome. Our data have also provided evidence that multiple-looping structure of long range interactions is a primary mechanism for transcription regulation by ER, which may facilitate the recruitment of collaborative co-factors and basal transcription components to the desired target sites.

These results convincingly demonstrated that ChIA-PET is an unbiased, *de novo*, and high-throughput approach for global analysis of long-range chromatin interactions mediated by protein factors. With the availability of the tag-based ultra high throughput sequencing technologies, ChIA-PET has the potential to become the most versatile and informative technology to map transcription interactomes mediated by all transcription factors and chromatin modifiers. Such whole genome binding and interaction information will certainly open a new field for understanding transcription regulation mechanisms at 3-demensional levels.

4.6 PERSPECTIVE

As exemplified in the above sections, PET technology has been well established as a primary technology for characterization of transcriptome and for elucidation of transcription regulatory networks. In addition, because of its unique nature for deriving the relationship between the two ends of test DNA fragments, it has been perceived to have broad potentials in applications to genomic analysis, such as for the analysis of genome structural variations (SVs) as well as for *de novo* individual genome shotgun sequencing and assembling.

Growing numbers of evidence have shown that human genomes undergone substantial structural variations including large and small pieces of insertion, inversion, translocation, and copy number variations (amplification or deletion). Traditional methods for detection karyotypic variations by array CGH and DNA FISH have low sensitivity and resolution. The combination

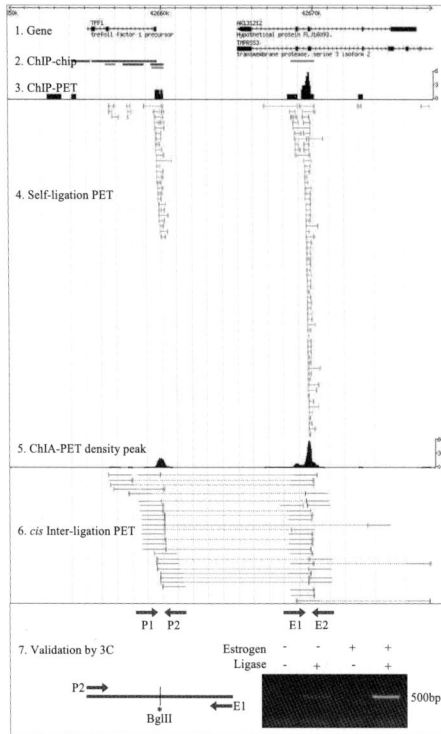

Figure 4.6. ER ChIA-PET data mapped at the TFF1 locus The ChIA-PET sequences were mapped to the reference genome for identification of ERα binding sites and ERα-mediated chromatin long-range interactions. The self-ligation PETs would indicate ERα binding site density, while the inter-ligation PETs would identify long-range interactions between two DNA fragments. The tracks of information included here are (starting from the top): **1.** UCSC Known Genes, the TFF1 gene locus. TFF1 is an estrogen-upregulated gene in MCF7 cells; **2.** ERα ChIP-chip (blue bars) **3.** ERá ChIP-PET density histogram shows ERα binding site peaks; **4.** Self-ligation ChIA-PET data that shows each PET with head and tail tags together with a solid horizontal line (orange) to represent the virtual ChIP DNA fragment. **5.** ERαChIA-PET density histogram showing the ERα binding sites in this region; **6.** Inter-ligation ChIA-PET data shows each tag with a vertical line and a solid horizontal line to represent the virtual DNA fragment from which the PET was derived from. A dotted line connects the paired tags of the same PET, indicating the interaction of the two DNA fragments. **7.** interactions identified by ChIA-PET data in this dataset were validated by ChIP-3C experiments. Negative controls of [estrogen –] and [ligation –] were included. Only [estrogen +] and [ligation +] reactions provided positive ChIP-3C products.

of PET analysis and the next generation sequencing has offered us with new promises for comprehensive genome structural variation analysis. Recent study has demonstrated such potential by using the 454 sequencing platform and paired end sequencing of 3 kb genomic DNA fragments from two human cell cultural lines[14] and identified extensive structural variations in the human genome. Due to abundant and large (>5 kb) repetitive regions scattered in the human genome, the current study with the ability of paired end sequencing covering only 2–3 kb genomic span would not be sufficient for comprehensive survey of human genome structural variations. It will be desirable to develop the capability of paired end sequencing cover genomic span >10 kb or higher.

Ultimately, with further improvement of the PET technology particularly in the robust preparation of PET libraries from large genomic DNA fragments (at 10 kb, 20 kb or 50 kb) and maturation of the ultra high throughput tag-based sequencing that could generate billions of tag reads per run, we foresee that the PET-based whole genome shotgun sequencing and assembling would become the method of choice for *de novo* personal human genome sequencing.

References

[1] A. Barski, S. Cuddapah, K. Cui, T. Y. Roh, D. E. Schones, Wang, Z. Wei, G. Chepelev, I. and K. Zhao. High-resolution profiling of histone methylations in the human genome. *Cell* **129**, 823–837 (2007).

[2] E. Birney, J. A. Stamatoyannopoulos, A. Dutta, R. Guigo, T. R. Gingeras, E. H. Margulies, Z. Weng, M. Snyder, E. T. Dermitzakis, R. E. Thurman *et al.* Identification and analysis of functional elements in 1% of the human genome by the ENCODE pilot project. *Nature* **447**, 799–816 (2007).

[3] L. A. Boyer, T. I. Lee, M. F. Cole, S. E. Johnstone, S. S. Levine, J. P. Zucker, M. G. Guenther, R. M. Kumar, H. L. Murray, R. G. Jenner *et al.* Core transcriptional regulatory circuitry in human embryonic stem cells. *Cell* **122**, 947–956 (2005).

[4] S. Brenner, M. Johnson, J. Bridgham, G. Golda, D. H. Lloyd, D. Johnson, S. Luo, S. McCurdy, M. Foy, M. Ewan *et al.* Gene expression analysis by massively parallel signature sequencing (MPSS) on microbead arrays. *Nat. Biotechnol.* **18**, 630–634 (2000).

[5] P. Carninci, and Y. Hayashizaki. High-efficiency full-length cDNA cloning. *Methods Enzymol.* **303**, 19–44 (1999).

[6] P. Carninci, T. Kasukawa, S. Katayama, J. Gough, M. C. Frith, N. Maeda, R. Oyama, T. Ravasi, B. Lenhard, C. Wells *et al*. The transcriptional landscape of the mammalian genome. *Science* **309**, 1559–1563 (2005).

[7] J. S. Carroll, C. A. Meyer, J. Song, W. Li, T. R. Geistlinger, J. Eeckhoute, A. S. Brodsky, E. K. Keeton, K. C. Fertuck, G. F. Hall *et al*. Genome-wide analysis of estrogen receptor binding sites. *Nat. Genet*. **38**, 1289–1297 (2006).

[8] D. Carter, L. Chakalova, C. S. Osborne, Y. F. Dai and P. Fraser. Long-range chromatin regulatory interactions *in vivo*. *Nat. Genet*. **32**, 623–626 (2002).

[9] J. Dekker, K. Rippe, M. Dekker and N. Kleckner. Capturing chromosome conformation. *Science* **295**, 1306–1311 (2002).

[10] J. Dostie, T. A. Richmond, R. A. Arnaout, R. R. Selzer, W. L. Lee, T. A. Honan, E. D. Rubio, A. Krumm, J. Lamb, C. Nusbaum *et al*. Chromosome Conformation Capture Carbon Copy (5C): A massively parallel solution for mapping interactions between genomic elements. *Genome Res*. **16**, 1299–1309 (2006).

[11] ENCODE (ENCyclopedia Of DNA Elements) *Project*. *Science* **306**, 636–640 (2004).

[12] P. Fraser and W. Bickmore. Nuclear organization of the genome and the potential for gene regulation. *Nature* **447**, 413–417 (2007).

[13] D. S. Johnson, A. Mortazavi, R. M. Myers and B. Wold. Genome-wide mapping of in vivo protein-DNA interactions. *Science* **316**, 1497–1502 (2007).

[14] J. O. Korbel, A. E. Urban, J. P. Affourtit, B. Godwin, F. Grubert, J. F. Simons, P. M. Kim, D. Palejev, N. J. Carriero, L. Du *et al*. Paired-end mapping reveals extensive structural variation in the human genome. *Science* **318**, 420–426 (2007).

[15] T. I. Lee, S. E. Johnstone and R. A. Young. Chromatin immunoprecipitation and microarray-based analysis of protein location. *Nat. Protoc*. **1**, 729–748 (2006).

[16] V. Levy, G. Sutton, P. C. Ng, L. Feuk, A. L. Halpern, B. P. Walenz, N. Axelrod, J. Huang, E. F. Kirkness, G. Denisov *et al*. The diploid genome sequence of an individual human. *PLoS Biol* **5**, e254 (2007).

[17] C. A. Lim, F. Yao, J. J. Wong, J. George, H. Xu, K. P. Chiu, W. K. Sung, L. Lipovich, V. B. Vega, J. Chen *et al*. Genome-wide mapping of RELA(p65) binding identifies E2F1 as a transcriptional activator recruited by NF-kappaB upon TLR4 activation. *Mol. Cell* **27**, 622–635 (2007).

[18] C. Y. Lin, V. B. Vega, J. S. Thomsen, T. Zhang, S. L. Kong, V. Xie, K. P. Chiu, L. Lipovich, D. H. Barnett, F. Stossi *et al*. Whole-genome cartography of estrogen receptor alpha binding sites. *PLoS Genet*. **3**, e87 (2007).

[19] Y. H. Loh, Q. Wu, J. L. Chew, V. B. Vega, W. Zhang, X. Chen, G. Bourque, J. George, B. Leong, J. Liu *et al*. The Oct4 and Nanog

transcription network regulates pluripotency in mouse embryonic stem cells. *Nat. Genet.* **38**, 431–440 (2006).

[20] M. Margulies, M. Egholm, W. E. Altman, S. Attiya, J. S. Bader, L. A. Bemben, J. Berka, M. S. Braverman, Y. J. Chen, Z. Chen *et al.* Genome sequencing in microfabricated high-density picolitre reactors. *Nature* **437**, 376–380 (2005).

[21] M. J. Mauro, M. O'Dwyer, M. C. Heinrich and B. J. Druker. STI571: a paradigm of new agents for cancer therapeutics. *J. Clin. Oncol.* **20**, 325–334 (2002).

[22] T. S. Mikkelsen, M. Ku, D. B. Jaffe, B. Issac, E. Lieberman, G. Giannoukos, P. Alvarez, W. Brockman, T. K. Kim, R. P. Koche *et al.* Genome-wide maps of chromatin state in pluripotent and lineage-committed cells. *Nature* **448**, 553–560 (2007).

[23] F. Mitelman, B. Johansson and F. Mertens. Fusion genes and rearranged genes as a linear function of chromosome aberrations in cancer. *Nat. Genet.* **36**, 331–334 (2004).

[24] P. Ng, J. J. Tan, H. S. Ooi, Y. L. Lee, K. P. Chiu, M. J. Fullwood, K. G. Srinivasan, C. Perbost, L. Du, W. K. Sung *et al.* Multiplex sequencing of paired-end ditags (MS-PET): A strategy for the ultra-high-throughput analysis of transcriptomes and genomes. *Nucleic Acids Res.* **34**, e84 (2006).

[25] P. Ng, C. L. Wei, W. K. Sung, K. P. Chiu, L. Lipovich, C. C. Ang, S. Gupta, A. Shahab, A. Ridwan, C. H. Wong *et al.* Gene Identification Signature (GIS) analysis for transcriptome characterization and genome annotation. *Nat. Methods* **2**, 105–111 (2005).

[26] C. S. Osborne, L. Chakalova, K. E. Brown, D. Carter, A. Horton, E. Debrand, B. Goyenechea, J. A. Mitchell, S. Lopes, W. Reik and P. Fraser. Active genes dynamically colocalize to shared sites of ongoing transcription. *Nat. Genet.* **36**, 1065–1071 (2004).

[27] Y. Ruan, H. S. Ooi, S. W. Choo, K. P. Chiu, X. D. Zhao, K. G. Srinivasan, F. Yao, C. Y. Choo, J. Liu, P. Ariyaratne *et al.* Fusion transcripts and transcribed retrotransposed loci discovered through comprehensive transcriptome analysis using Paired-End diTags (PETs). *Genome Res.* **17**, 828–838 (2007).

[28] S. Saha, A. B. Sparks, C. Rago, V. Akmaev, C. J. Wang, B. Vogelstein, K. W. Kinzler and V. E. Velculescu. Using the transcriptome to annotate the genome. *Nat. Biotechnol.* **20**, 508–512 (2002).

[29] J. Shendure, G. J. Porreca, N. B. Reppas X. Lin, J. P. McCutcheon, A. M. Rosenbaum, M. D. Wang, K. Zhang, R. D. Mitra and G. M. Church. Accurate multiplex polony sequencing of an evolved bacterial genome. *Science* **309**, 1728–1732 (2005).

[30] D. D. Shoemaker, E. E. Schadt, C. D. Armour, Y. D. He, P. Garrett-Engele, P. D. McDonagh, P. M. Loerch, A. Leonardson, P. Y. Lum, G. Cavet *et al.* Experimental annotation of the human genome using microarray technology. *Nature* **409**, 922–927 (2001).

[31] C. G. Spilianakis, M. D. Lalioti, T. Town, G. R. Lee and R. A. Flavell. Interchromosomal associations between alternatively expressed loci. *Nature* **435**, 637–645 (2005).

[32] R. L. Strausberg, E. A. Feingold, L. H. Grouse, J. G. Derge, R. D. Klausner, F. S. Collins, L. Wagner, C. M. Shenmen, G. D. Schuler, S. F. Altschul *et al.* Generation and initial analysis of more than 15,000 full-length human and mouse cDNA sequences. *Proc. Natl. Acad. Sci. USA* **99**, 16899–16903 (2002).

[33] V. E. Velculescu, L. Zhang, B. Vogelstein and K. W. Kinzler. Serial analysis of gene expression. *Science* **270**, 484–487 (1995).

[34] C. L. Wei, P. Ng, K. P. Chiu, C. H. Wong, C. C. Ang, L. Lipovich, E. T. Liu and Y. Ruan. 5′ Long serial analysis of gene expression (LongSAGE) and 3′ LongSAGE for transcriptome characterization and genome annotation. *Proc. Natl. Acad. Sci. USA* **101**, 11701–11706 (2004).

[35] C. L. Wei, Q. Wu, V. B. Vega, K. P. Chiu, P. Ng, T. Zhang, A. Shahab, H. C. Yong, Y. Fu, Z. Weng *et al.* A global map of p53 transcription-factor binding sites in the human genome. *Cell* **124**, 207–219 (2006).

[36] J. Wu, L. T. Smith, C. Plass and T. H. Huang. ChIP-chip comes of age for genome-wide functional analysis. *Cancer Res.* **66**, 6899–6902 (2006).

[37] K. I. Zeller, X. Zhao, C. W. Lee, K. P. Chiu, F. Yao, J. T. Yustein, H. S. Ooi, Y. L. Orlov, A. Shahab, H. C. Yong *et al.* Global mapping of c-Myc binding sites and target gene networks in human B cells. *Proc. Natl. Acad. Sci. USA* **103**, 17834–17839 (2006).

[38] X. D. Zhao, X. Han, J. L. Chew, J. Liu, K. P. Chiu, A. Choo, Y. L. Orlov, W. K. Sung, A. Shahab, V. A. Kuznetsov *et al.* Whole-genome mapping of histone h3 lys4 and 27 trimethylations reveals distinct genomic compartments in human embryonic stem cells. *Stem Cell* **1**, 286–298 (2007).

[39] Z. Zhao, G. Tavoosidana, M. Sjolinder, A. Gondor, P. Mariano, S. Wang, C. Kanduri, M. Lezcano, K. S. Sandhu, U. Singh *et al.* Circular chromosome conformation capture (4C) uncovers extensive networks of epigenetically regulated intra- and interchromosomal interactions. *Nat. Genet.* **38**, 1341–1347 (2006).

[40] F. S. Collins, E. S. Lander, J Rogers and R. H. Waterson. Finishing the euchromatic sequence of the human genome. *Nature* **431**(7011), 931–945 (2004).

Chapter Five

New Era of Genome-Wide Gene Expression Analysis

Norihiro Maeda

Genome Science Laboratory, Discovery and Research Institute,
RIKEN Wako Main Campus, Japan
and
Omics Sciences Center, RIKEN Yokohama Institute, Japan
Email: norihiro@riken.jp

Hundreds of genome sequences are now open to the public, but the information on gene function encoded by the genome remains limited. Two approaches, microarray-based and tag-based, are mainly applied for comprehensive understanding of gene function. The microarray-based approach provides us easy access to genome-wide analysis, but the detection level is limited by the set of probes on the chip. In contrast, the tag-based approach offers an open system, allowing us to survey gene expression genome-wide without specifying genes of interest. Additionally, the tag-based approach attracts the attention of scientists because it provides the exact positions (genomic loci) where genes are transcribed. However, conventional sequencing technology like Sanger method requires labour-intensive procedures, hindering the potential of tagging technology to be realized. The recent advent of next generation sequencing technologies revolutionizes the field of genome-wide gene expression analysis. Next generation sequencers are capable of sequencing at the level of gigabases within a week or a few weeks, leading us to a new era of genome-wide gene expression analysis. Here, I describe the principles of next generation sequencing technologies and discuss the application of these technologies to genome-wide analysis.

Cap Analysis Gene Expression (CAGE): The Science of Decoding Gene
Transcription **edited by P Carninci**
Copyright © 2010 by Pan Stanford Publishing Pte Ltd
www.panstanford.com
978-981-4241-34-2

5.1 INTRODUCTION

Since 1990's, microarrays have been widely recognized as powerful tools to measure the level of gene expression.[1,2] Microarray-based analysis is routinely used because it requires relatively simple sample preparation, but it is difficult to control unintended cross hybridization. The high GC content of probes on microarray chip tends to cause cross hybridization.[3] It is also reported that controversial conclusions can be drawn from comparison between analyses/platforms in some cases.[4,5] In addition, the detection level is limited by the set of probes on the chip unless one uses genome-wide tiling arrays. The microarray-based analysis is called a closed system.

In contrast, the tag-based approach, like the types discussed in this essay, is called an open system and it allows us global gene expression analysis without narrowing down the scope of research interest. To date, several tagging technologies such as CAGE (Cap Analysis of Gene Expression),[6,7] MS-PET (Multiplex Sequencing of Paired-End Ditags)/GIS-PET (Gene Identification Signature; see also Chapter 4),[8,9] SAGE (Serial Analysis of Gene Expression; seel also chapter 1),[9−11] and MPSS (Massively Parallel Signature Sequencing)[12] are widely recognized. The advantage of tag-based approaches is that they are able to tell us unambiguous positions of transcripts through mapping tags to a respective genome. Additionally, the number of sequenced tags reflects the level of gene expression. Thus, tagging technologies have great potential, but the sequencing capacity of Sanger method had so far limited the potential scale of this approach.

Recently, several next generation sequencers became commercially available and these sequencers are able to sequence billion or sub-billion bases in a few weeks, promising to lead us to a new era of genome-wide analysis. In this chapter, the principles of next generation sequencers are introduced, and potential applications of tagging technologies with next generation technologies are discussed.

5.2 TAGGING TECHNOLOGIES FOR GENOME-WIDE ANALYSIS

To understand an organism comprehensively, gene expression analysis provides us a clue as to which genes are transcribed in which cells/tissues. To address this question, several tag-based approaches for identifying transcripts have been so far developed. Compared to full-length sequencing of transcripts, the tagging technologies give us more information on transcripts with small number of sequencing paths because the sequences of tags extracted from several transcripts can be determined with a single sequencing run. The tagging approaches have been recognized as a cost-effective means to discover rare transcripts and/or to promoter analysis.

CAGE is a method developed to extract a short sequence from the 5′ end of a transcript.[6,7] Twenty or twenty-one nucleotides are extracted from cap-trapped transcripts by *Mme*I enzyme (see Chapter 2 for an outline and Chapter 3 for experimental details). This CAGE method was applied previously in a genome-wide gene expression analysis and promoter analysis.[13,14] MS-PET/GIS-PET technologies provide us short sequences of both 5′ and 3′ ends of a transcript simultaneously.[8,9] This PET method also contributes to identifying DNA-binding sites for transcription factors, for instance, with combination of chromatin immunoprecipitation (ChIP).[9] SAGE was firstly reported in 1995 as a method to compare gene expression profiling between normal and disease states.[15] At the beginning, the SAGE method extracted ca. 13 nucleotides as a tag, but the SAGE was further developed by exchanging restriction endonuclease (*Mme*I, *Eco*P15I) to extract longer nucleotides as tags (LongSAGE, SuperSAGE).[10,11] Longer tags, 21 bp in the LongSAGE and 26 bp in the SuperSAGE, improved the efficiency of tag-to-gene annotation. Another tagging technology, now not in use any longer, is called MPSS, and this technology adopts *in vitro* cloning of DNA templates on microbeads. The sequences of tags are obtained by confocal fluorescent microscope through multi cycles of ligation-cleavage using type II restriction endonuclease.[12]

5.3 PRINCIPLES OF NEXT GENERATION SEQUENCING TECHNOLOGIES

5.3.1 *Genome Sequencer 20/FLX System (Roche Diagnostics/454 Life Sciences) (See Figure 5.1)*

The principle of Genome sequencer 20 (GS 20)/Genome Sequencer FLX (GS FLX) is based on emulsion PCR using nano-scale beads and pyrosequencing.[16] One instrumental run of GS 20

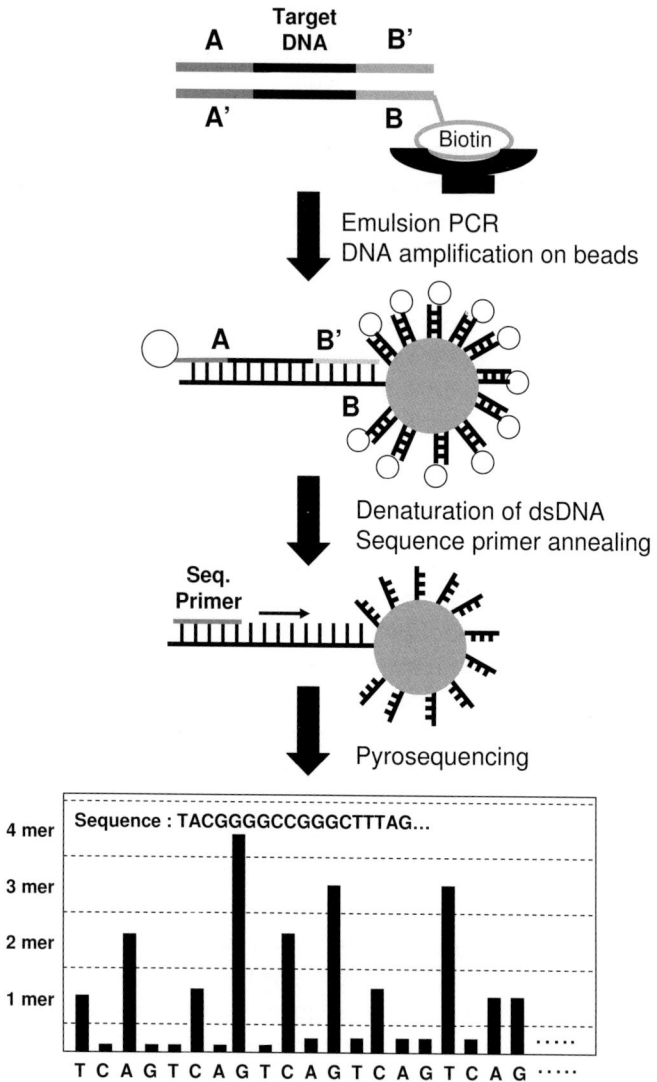

Figure 5.1. Principle of the Genome Sequencer 20/FLX system.

provides more than 200,000 reads with about 100 bases average in length, resulting in producing 20 million bases. In the case of GS FLX, which is upgraded version of GS 20, single run produces greater than 400,000 reads ranging 200–300 bases on average, while the new Titanium promises over 1 million reads that are 400 nt long. The GS 20 was initially developed by 454 life sciences for sequencing bacterial genomes as one of main applications. Since commercial availability, the potential of this sequence technology has been realized to the fields of comparative genomics, small RNAs, gene regulation, epigenetics, metagenomics, transcriptome, paleogenomics, and pharmacogenomics.[17-30] More than 100 research articles using GS 20 have been already reported.

For GS 20/GS FLX sequencing, double-stranded DNAs (dsDNA) are required as starting materials. In the case of tagging technology for gene expression analysis, the length of each tag is short, 20 bases, and concatenation of tags by ligation is necessary to achieve the full potential of the sequencer. Double-stranded DNAs are then ligated to sequencer-specific adaptors (A and B) at the both ends. The weak point of tagging technology is the limited amount of DNAs because only ca. 20 bases can be extracted from each transcript. As for ditag technology, the length of extracted tags should be double, but the quantity of extracted DNA is still limited. Therefore, PCR amplification during sample preparation procedures is indispensable for obtaining enough quantity of DNA for sequencing. The GS 20/GS FLX basically requires dsDNA in the order of about one to few micrograms. If necessary, the sequences of sequencer-specific adaptors should be introduced upon designing amplification primers to reduce the loss of DNA during experimental manipulation. Following the preparation of dsDNA, single-stranded template DNA (sstDNA) is extracted from adaptor-ligated dsDNA by using biotin attached to one of sequencer-specific adaptors. Extracted sstDNA is subjected to emulsion PCR together with amplification beads. Following emulsion PCR amplification, the beads on which DNA is successfully amplified are selected by using biotin which is incorporated through emulsion PCR. The resulting beads are applied to wells of PicoTiterPlate. On the PicoTiterPlate, pyrosequencing is performed.[31] The signal intensity by each nucleotide flow is converted into the number of incorporated nucleotides.

5.4 GENOME ANALYZER (ILLUMINA/SOLEXA)

Genome analyzer works based on sequencing-by-synthesis with four fluorescently-labeled modified nucleotides.[32] This sequencing technology was originally developed by Solexa, Inc, and it was taken over by Illumina, Inc. One instrumental run of Illumina genome analyzer produces more than one billion bases. This technology has been applied to not only genome (re)-sequencing, but also ChIP-seq analysis, gene expression, epigenetic analysis, and small RNA profiling.[33-38] Users are able to choose reagents from three sequencing kits for sequencing 17, 25,

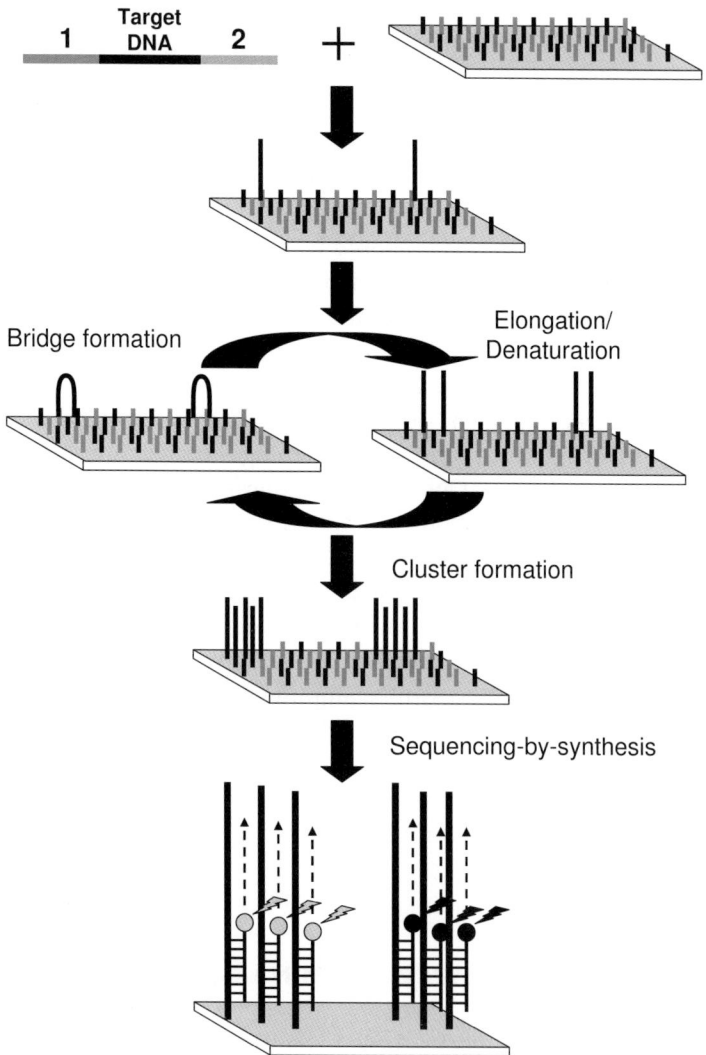

Figure 5.2. Principle of the Genome Analyzer.

or 35 bases par read, but this is a moving target because pair end reads 75 nt longer will be launched soon.

Contrary to the other two sequencing technologies, the Illumina genome analyzer does adopt not emulsion PCR but "bridge amplification". Following attaching sequencer-specific adaptors (1 and 2) to sample "target" DNAs at the both ends, the DNAs are denatured with alkali (NaOH) and immobilized at one end on the plate, whose surface is densely covered with adaptors, through hybridization. With keeping this hybridization, the other end of ssDNA is supposed to reach a nearby complementary adaptor which is attached to the surface of the plate, resulting in formation a bridge. Using this bridge, complementary strand is formed with unlabeled nucleotides by polymerase, forming double-stranded bridge. In this procedure, a single-stranded DNA bridge and the nearby adaptor work as a template and a primer, respectively. The dsDNA bridge is then denatured and each strand is subjected to form hybridization with another nearby adaptor on the plate, resulting in forming other bridges. This bridge formation-amplification process is iterated and at the end, around 1000 identical copies of each single DNA are created in close proximity to each other. In total, several million dense clusters of identical dsDNA are generated in each channel of the flow cell. Following the cluster formation, all four labeled reversible terminators, primers, and DNA polymerase are applied to the plate for sequencing. The incorporated nucleotides are excited by laser, and the resultant emitted fluorescence from each cluster is recorded to determine the nucleotide. After washing of unincorporated nucleotides, primers, and enzymes, the sequencing cycle is repeated. When the sequencing cycle completes, the DNA sequence of each cluster is determined using the combination of the order of emitted fluorescence and the information about the coordinate where fluorescence is emitted.

5.5 SOLID SYSTEM (APPLIED BIOSYSTEMS)

The unique characteristic of SOLiD system is that this sequencing system is based on sequential ligation by DNA ligase (not elongation by polymerase).[39] Fundamental technology for SOLiD system was initially developed by Agencourt Personal Genomics, and Applied Biosystems further developed and commercialized

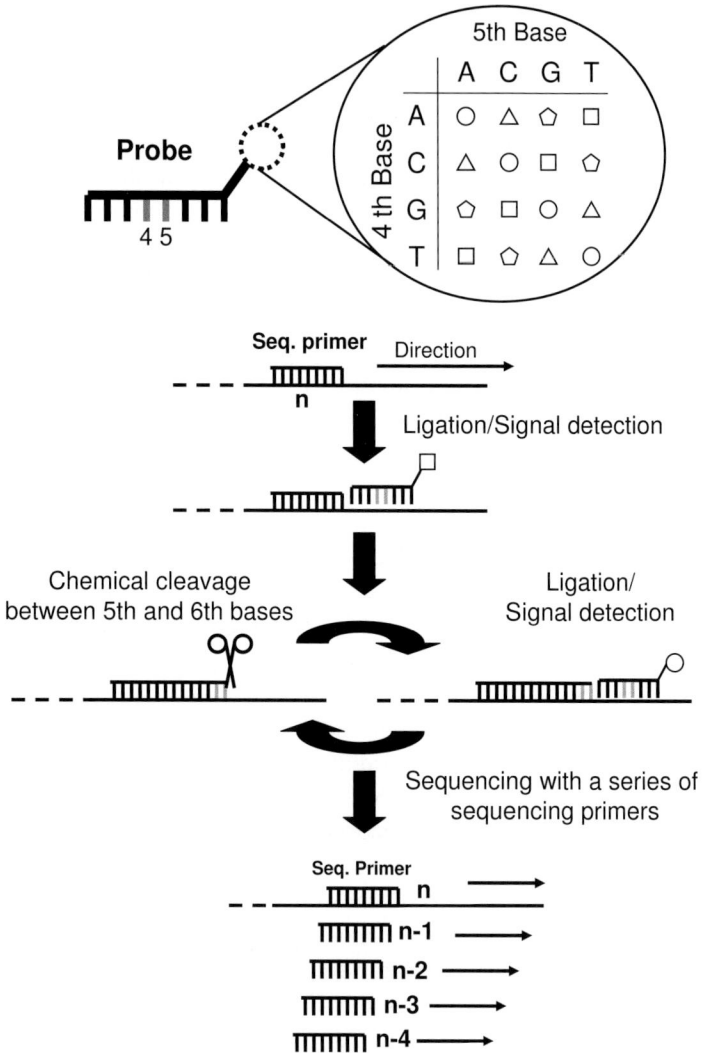

Figure 5.3. Principle of the SOLiD system.

the SOLiD system (see Figure 5.3). The SOLiD system adopts a two base encoding technology to obtain high raw base accuracy, which is a very attractive feature. A single instrument run is capable of determining more than one billion bases. The read length is 25 or 35 bases.

Double-stranded DNAs are first ligated with sequencer-specific DNA adaptors (P1, P2) at the both ends. The ligated DNAs are amplified on the surface of beads by emulsion PCR. The amplified beads are deposited to open glass slide for sequencing and are fixed through covalent bonding by 3′ modification of amplified DNA. During sequencing procedures, sequencing primers (Position: n) are hybridized to the adapter sequence as a first step. And then, a set of four color dye-labeled probes compete for ligation to the sequencing primer. Each probe is eight bases in length, and the color of dye attached to a probe depends on the combination of 4th and 5th bases out of 8 bases. After detecting the color of ligated probe, the ligated probe is chemically cleaved between 5th and 6th bases. The 3′ part of cleaved probe (6th through 8th bases) is washed away, and a next ligation cycle takes place. This ligation-dye detection-cleavage procedure is repeated, determining 2 nucleotides at the interval of 5 bases. Following the sequential sequence determination, newly synthesized strand is denatured and washed away. Next, another sequencing primer, whose sequence is shifted by one nucleotide (Position: n-1), is annealed to the adaptor sequence, and the ligation-dye detection-cleavage procedure is repeated as above in order to determine 2 nucleotides at the interval of 5 bases. The series of detection procedures are reiterated 5 times (Positions: n, n-1, n-2, n-3, n-4, and n-5). Therefore, every nucleotide on target DNA is base-called twice by two series of detection procedures, resulting in increasing the sequencing accuracy.

5.6 ADVANTAGES OF NEXT GENERATION SEQUENCING TECHNOLOGIES OVER CONVENTIONAL SEQUENCING TECHNOLOGY ON TAGGING TECHNOLOGIES

Apparently, one of the big advantages of next generation sequencing technologies is their huge throughput. One single instrumental run generates more than 100 Mb, and some technologies are capable of producing more than one billion bases per run. Capillary sequencing based on Sanger method gives 600–700 bases per read, but it is not comparable to next generation technologies in terms of throughput. Additionally, the conventional sequencing technology requires us laborious procedures such as cloning, cultivating *E. coli* harboring plasmids, colony picking, extracting and purifying plasmid DNAs. For performing a transcriptome project, for instance, more than millions of clones and plasmids must be manipulated. To do this, a number of operators must work intensively for months handling thousands of plates. In contrast, next generation sequencing technologies liberate us from such laborious steps and let us go sequencing directly with dsDNAs as initial materials. Clonal amplification of DNAs for sequencing can be carried out in parallel in a few tubes/plates or on a sequencing plate. Thus, a single operator is now able to generate more than million/billion bases in a week using next generation technologies.

To date, more than 500 genome sequences have been determined and more than 1000 genome sequencing projects are underway or published their draft assembly. The number of genome sequencing projects is expected to increase dramatically in near future. The availability of more genome sequences largely helps us mapping the tags extracted from transcripts to the reference genomes, accelerating to broaden the fields of tagging technology application.

It should be noted that read length of 25 nucleotides give more than 75% of uniquely mapping to a human genome, and that ca. 60 nucleotides is required for 95% uniqueness.[40] Read lengths of next generation sequencing technologies range around 200–300 nucleotides (GS FLX) and 20–35 nucleotides (Genome Analyzer/SOLiD system). In addition, the restriction endonuclease which gives 27 bp DNA fragments is available. Altogether, the current tag-based approach combined with next generation sequencing technologies can cover more than four-thirds of the human genome.

5.7   FROM STATIC ANALYSIS TO DYNAMIC ANALYSIS

Due to the limitation of throughput, the genome-wide analyses using tagging technologies has been restricted in the past to static analyses such as comparison between normal and diseased cells and investigation on cell-specific gene expression.[14] The recent advent of next generation sequencing technologies dramatically reduces sequencing cost and time, leading us to challenge for an unexplored field of global gene expression analysis. One of next challenges is to monitor dynamic changes in gene expression. Recently, the data on protein-protein interaction has been accumulated at high speed. A number of studies drawing the network based on protein-protein interaction analysis have been reported.[41-43] However, one more axis, time-course, is necessary for understanding gene function comprehensively. For instance, the time-course analysis is indispensable to understand the mechanisms/pathways underlying proliferation, differentiation, cellular response to external stimulus and so on.

To address this question, a series of RNA should be extracted from cells at various time points and the tags obtained should be comparable between time-course samples. One of advantages of next generation sequencing technologies is that double-stranded DNA can be utilized as starting material, but the disadvantage of these technologies is that they require the quantity of dsDNA at the level of microgram or sub-microgram. The tag-based approach extracting a 20–30 nucleotide tag from each transcript faces the issue on the amount of DNA, and PCR amplification of tags is indispensable for sequencing. However, PCR amplification during sample preparation has the potential to cause amplification bias. Especially, the issue about amplification bias should be critical in the case that tags from each RNA source are generated independently.

One of options to solve the problem on quantity is to mix RNA samples by introducing "DNA barcode" during tag preparation. The DNA barcode composes of short nucleotides, for example, 4 or 5 nucleotides, and each DNA barcode corresponds to each RNA source. With introducing DNA barcodes, tags from diverse RNA sources can be pooled at the early stage of sample preparation, resulting in increasing the quantity of starting material. It is apparent that the increased amount of DNA contributes to reducing the number of PCR cycles, suggesting that amplification bias can be suppressed. In addition, this pooling approach allows us to prepare samples under uniform conditions, leading to producing comparable dataset. The pooled approach is also favorable from a technological viewpoint. When the sequencing samples are prepared independently, the optimal conditions for sequencing should be established for each sample. In the case of the pooled approach, once the optimal conditions for sequencing are obtained, the sequencing procedures should be repeated routinely. Thus, the pooled approach helps saving time and budget.

Assumed that one instrumental run determines 1 billon bases and each tag has 30 bases in length, more than 30 million tags per run are theoretically produced. In a previous study, we used ca. 5 million human CAGE tags from 41 different human libraries or ca. 7 million mouse CAGE tags from 145 different mouse libraries for promoter analysis.[14] Taken together, the throughput of next generation sequencers seems to be sufficient for monitoring dynamic changes in gene expression.

5.8 CAGE METHOD AND NEXT GENERATION SEQUENCING TECHNOLOGIES

To explore global dynamic expression changes, the throughput of sequencer which affects the outcome is a key factor to select a sequencer. In terms of the throughput, the genome analyzer and the SOLiD system, which are capable of producing more than one billion bases par instrumental run, should be fist choices for the moment. However, it should be noted that these two technologies generate short reads such as 35 nucleotides. From a historical viewpoint, tagging technologies are moving forward to extract longer tags from transcripts through replacing restriction endonucleases in order to improve the mapping rate of extracted tags to reference genomes.[10,11,15] In some tagging technologies like SAGE, short tags from transcripts were ligated tail-to-tail, resulting in forming ditags to simplify the experimental protocols. Therefore, the read length itself potentially causes obstacles upon modifying experimental protocols for next generation sequencers.

In the case of the CAGE method, 20 or 21 nucleotides (26–27 bp when used *Eco*P15I) are extracted from the 5′-end of a transcript.[6] Even 4 or 5 base-DNA barcode is employed, the entire size of tags in length is within the scope of next generation sequencers, indicating that the CAGE method is relatively flexible for adapting new sequencing technologies. This flexibility is one of advantages for the CAGE method. Theoretically, four or five base-DNA barcode provides 256 or 1024 sorts of barcodes. Even some problematic barcodes such as the barcodes harboring homopolymers in the middle are excluded, the variation of barcodes is enough to identify RNA sources/libraries. In addition, we should carefully consider the complexity of barcodes upon using next generation sequencers. If only a few barcodes are employed, it means that a huge number of sample DNAs give signals simultaneously during sequencing steps, causing the interference between signals and/or the lack of reagents. Therefore, proper selection and/or combination of barcodes are required. In some cases, several barcodes should be employed to identify one RNA source/library.

As above, next generation sequencing technologies require us to ligate sequencer-specific adaptors to the both ends of sample DNAs. The CAGE method takes a cap-trapping strategy, and it is quite easy to introduce barcodes and sequencer-specific adaptors upstream of CAGE tags by changing the sequences of CAGE method-specific linkers. In addition, the CAGE method provides exact transcription start sites, allowing us intensive promoter analysis. From this point of view, the CAGE method has advantages.

5.9 CONCLUSIONS AND OUTLOOK

The advent of next generation sequencing technologies induced drastic changes in genome-wide analysis. It is safe to say that these techniques successfully dropped the sequencing cost by orders of magnitude, and that they change the research style for genome-wide analysis. For instance, a microarray-based ChIP-Chip technique has been replaced by a sequencer-based ChIP-Seq. Millions of small RNAs can be sequenced within a week or a few weeks. Draft sequence of a bacterial genome can be determined within a week. We are stepping into a new phase of genome-wide analysis.

In this chapter, three technologies are introduced, but the development of next generation sequencing technologies is making progress very rapidly and novel sequencing technologies at single molecule level are now under development. Taken into account the fact that we are still in the middle of $1000 genome project, the circumstances of next generation sequencing technologies are continuing to change drastically in a short period of time. At the publication of this book, some other new technologies might be commercially available.

Existing sequencing technologies are also being further improved. Read length and total number of reads of next generation sequencers are being planned to be lengthened and increased, respectively. We are reaching the era where a single instrumental run generates sequenced bases equal to the bases of our genome. Concrete collaboration and coordination between "wet" and "dry" scientists are crucial for comprehensive understanding of gene network.

ACKNOWLEDGMENTS

I thank the members of the GERG and GSL RIKEN for their support. This work was supported by a Research Grant for National Project on Protein Structural and Functional Analysis from MEXT and a grant of the Genome Network Project from the Ministry of Education, Culture, Sports, Science and Technology, Japan.

References

[1] J. Cheng, P. Kapranov, J. Drenkow, S. Dike, S. Brubaker, S. Patel, J. Long, D. Stern, H. Tammana, G. Helt, V. Sementchenko, A. Piccolboni, S. Bekiranov, D. K. Bailey, M. Ganesh, S. Ghosh, I. Bell, D. S. Gerhard and T. R. Gingeras. *Science* **308**, 1149 (2005).

[2] P. Kapranov, J. Cheng, S. Dike, D. A. Nix, R. Duttagupta, A. T. Willingham, P. F. Stadler, J. Hertel, J. Hackermuller, I. L. Hofacker, I. Bell, E. Cheung, J. Drenkow, E. Dumais, S. Patel, G. Helt, M. Ganesh, S. Ghosh, A. Piccolboni, V. Sementchenko *et al. Science* **316**, 1484 (2007).

[3] D. Sasaki, S. Kondo, N. Maeda, T. R. Gingeras, Y. Hasegawa and Y. Hayashizaki. *Genomics* **89**, 541 (2007).

[4] S. Draghici, P. Khatri, A. C. Eklund and Z. Szallasi. *Trends Genet.* **22**, 101 (2006).

[5] J. M. Johnson, S. Edwards, D. Shoemaker and E. E. Schadt. *Trends Genet.* **21**, 93 (2005).

[6] T. Shiraki, S. Kondo, S. Katayama, K. Waki, T. Kasukawa, H. Kawaji, R. Kodzius, A. Watahiki, M. Nakamura, T. Arakawa, S. Fukuda, D. Sasaki, A. Podhajska, M. Harbers, J. Kawai, P. Carninci and Y. Hayashizaki. *Proc. Natl. Acad. Sci. USA* **100**, 15776 (2003).

[7] R. Kodzius, M. Kojima, H. Nishiyori, M. Nakamura, S. Fukuda, M. Tagami, D. Sasaki, K. Imamura, C. Kai, M. Harbers, Y. Hayashizaki and P. Carninci. *Nat. Methods* **3**, 211 (2006).

[8] P. Ng, C. L. Wei, W. K. Sung, K. P. Chiu, L. Lipovich, C. C. Ang, S. Gupta, A. Shahab, A. Ridwan, C. H. Wong, E. T. Liu and Y. Ruan. *Nat. Methods* **2**, 105 (2005).

[9] P. Ng, J. J. Tan, H. S. Ooi, Y. L. Lee, K. P. Chiu, M. J. Fullwood, K. G. Srinivasan, C. Perbost, L. Du, W. K. Sung, C. L. Wei and Y. Ruan, *Nucleic Acids Res.* **34**, e84 (2006).

[10] S. Saha, A. B. Sparks, C. Rago, V. Akmaev, C. J. Wang, B. Vogelstein, K. W. Kinzler and V. E. Velculescu. *Nat. Biotechnol.* **20**, 508 (2002).

[11] H. Matsumura, S. Reich, A. Ito, H. Saitoh, S. Kamoun, P. Winter, G. Kahl, M. Reuter, D. H. Kruger and R. Terauchi. *Proc. Natl. Acad. Sci. USA* **100**, 15718 (2003).

[12] S. Brenner, M. Johnson, J. Bridgham, G. Golda, D. H. Lloyd, D. Johnson, S. Luo, S. McCurdy, M. Foy, M. Ewan, R. Roth, D. George, S. Eletr, G. Albrecht, E. Vermaas, S. R. Williams, K. Moon, T. Burcham, M. Pallas, R. B. DuBridge *et al. Nat. Biotechnol.* **18**, 630 (2000).

[13] P. Carninci, T. Kasukawa, S. Katayama, J. Gough, M. C. Frith, N. Maeda, R. Oyama, T. Ravasi, B. Lenhard, C. Wells, R. Kodzius, K. Shimokawa, V. B. Bajic, S. E. Brenner, S. Batalov, A. R. Forrest, M. Zavolan, M. J. Davis, L. G. Wilming, V. Aidinis *et al. Science* **309**, 1559 (2005).

[14] P. Carninci, A. Sandelin, B. Lenhard, S. Katayama, K. Shimokawa, J. Ponjavic, C. A. Semple, M. S. Taylor, P. G. Engstrom, M. C. Frith, A. R. Forrest, W. B. Alkema, S. L. Tan, C. Plessy, R. Kodzius, T. Ravasi, T. Kasukawa, S. Fukuda, M. Kanamori-Katayama, Y. Kitazume *et al. Nat. Genet.* **38**, 626 (2006).

[15] V. E. Velculescu, L. Zhang, B. Vogelstein and K. W. Kinzler. *Science* **270**, 484 (1995).

[16] M. Margulies, M. Egholm, W. E. Altman, S. Attiya, J. S. Bader, L. A. Bemben, J. Berka, M. S. Braverman, Y. J. Chen, Z. Chen, S. B. Dewell, L. Du, J. M. Fierro, X. V. Gomes, B. C. Godwin, W. He, S. Helgesen, C. H. Ho, G. P. Irzyk, S. C. Jando *et al. Nature* **437**, 376 (2005).

[17] K. H. Taylor, R. S. Kramer, J. W. Davis, J. Guo, D. J. Duff, D. Xu, C. W. Caldwell and H. Shi. *Cancer Res.* **67**, 8511 (2007).

[18] K. Andries, P. Verhasselt, J. Guillemont, H. W. Gohlmann, J. M. Neefs, H. Winkler, J. Van Gestel, P. Timmerman, M. Zhu, E. Lee, P. Williams, D. de Chaffoy, E. Huitric, S. Hoffner, E. Cambau, C. Truffot-Pernot, N. Lounis and V. Jarlier. *Science* **307**, 223 (2005).

[19] G. J. Velicer, G. Raddatz, H. Keller, S. Deiss, C. Lanz, I. Dinkelacker and S. C. Schuster. *Proc. Natl. Acad. Sci. USA* **103**, 8107 (2006).

[20] K. L. Nielsen, A. L. Hogh and J. Emmersen. *Nucleic Acids Res.* **34** e133 (2006).

[21] A. L. Toth, K. Varala, T. C. Newman, F. E. Miguez, S. K. Hutchison, D. A. Willoughby, J. F. Simons, M. Egholm, J. H. Hunt, M. E. Hudson and G. E. Robinson. *Science* **318**, 441 (2007).

[22] S. Leininger, T. Urich, M. . Schloter, L. Schwark, J. Qi, G. W. Nicol, J. I. Prosser, S. C. Schuster and C. Schleper. *Nature* **442**, 806 (2006).

[23] J. A. Huber, D. B. Mark Welch, H. G. Morrison, S. M. Huse, P. R. Neal, D. A. Butterfield and M. L. Sogin. *Science* **318**, 97 (2007).

[24] I. R. Henderson, X. . Zhang, C. Lu, L. Johnson, B. C. Meyers, P. J. Green and S. E. Jacobsen. *Nat. Genet.* **38**, 721 (2006).

[25] A. Girard, R. Sachidanandam, G. J. Hannon and M. A. Carmell. *Nature* **442**, 199 (2006).

[26] Y. Qi, X. He, X. J. Wang, O. Kohany, J. Jurka and G. J. Hannon. *Nature* **443**, 1008 (2006).

[27] H. N. Poinar, C. Schwarz, J. Qi, B. Shapiro, R. D. Macphee, B. Buigues, A. Tikhonov, D. H. Huson, L. P. Tomsho, A. Auch, M. Rampp, W. Miller and S. C. Schuster. *Science* **311**, 392 (2006).

[28] R. E. Green, J. Krause, S. E. Ptak, A. W. Briggs, M. T. Ronan, J. F. Simons, L. Du, M. Egholm, J. M. Rothberg, M. Paunovic and S. Paabo. *Nature* **444**, 330 (2006).

[29] J. P. Noonan, G. Coop, S. Kudaravalli, D. Smith, J. Krause, J. Alessi, F. Chen, D. Platt, S. Paabo, J. K. Pritchard and E. M. Rubin. *Science* **314**, 1113 (2006).

[30] M. T. Gilbert, L. P. Tomsho, S. Rendulic, M. Packard, D. I. Drautz, A. Sher, A. Tikhonov, L. Dalen, T. Kuznetsova, P. Kosintsev, P. F. Campos, T. Higham, M. J. Collins, A. S. Wilson, F. Shidlovskiy, B. Buigues, P. G. Ericson, M. Germonpre, A. Gotherstrom, P. Iacumin *et al. Science* **317**, 1927 (2007).

[31] M. Ronaghi. *Genome Res.* **11**, 3 (2001).

[32] D. R. Bentley. *Curr. Opin. Genet. Dev.* **16**, 545 (2006).

[33] A. Barski, S. Cuddapah, K. Cui, T. Y. Roh, D. E. Schones, Z. Wang, G. Wei, I. Chepelev and K. Zhao. *Cell* **129**, 823 (2007).

[34] D. S. Johnson, A. Mortazavi, R. M. Myers and B. Wold. *Science* **316**, 1497 (2007).

[35] G. Robertson, M. Hirst, M. Bainbridge, M. Bilenky, Y. Zhao, T. Zeng, G. Euskirchen, B. Bernier, R. Varhol, A. Delaney, N. Thiessen, O. L. Griffith, A. He, M. Marra, M. Snyder and S. Jones. *Nat. Methods* **4**, 651 (2007).

[36] T. S. Mikkelsen, M. Ku, D. B. Jaffe, B. Issac, E. Lieberman, G. Giannoukos, P. Alvarez, W. Brockman, T. K. Kim, R. P. Koche, W. Lee, E. Mendenhall, A. O'Donovan, A. Presser, C. Russ, X. Xie, A. Meissner, M. Wernig, R. Jaenisch, C. Nusbaum *et al. Nature* **448**, 553 (2007).

[37] E. Hodges, Z. Xuan, V. Balija, M. Kramer, M. N. Molla, S. W. Smith, C. M. Middle, M. J. Rodesch, T. J. Albert, G. J. Hannon and W. R. McCombie. *Nat. Genet* **39**, 1522 (2007).

[38] G. J. Porreca, K. Zhang, J. B. Li, B. Xie, D. Austin, S. L. Vassallo, E. M. LeProust, B. J. Peck, C. J. Emig, F. Dahl, Y. Gao, G. M. Church and J. Shendure. *Nat. Methods* **4**, 931 (2007).

[39] J. Shendure, G. J. Porreca, N. B. Reppas, X. Lin, J. P. McCutcheon, A. M. Rosenbaum, M. D. Wang, K. Zhang, R. D. Mitra and G. M. Church. *Science* **309**, 1728 (2005).

[40] N. Whiteford, N. Haslam, G. Weber, A. Prugel-Bennett, J. W. Essex, P. L. Roach, M. Bradley and C. Neylon. *Nucleic Acids Res.* **33**, e171 (2005).

[41] J. F. Rual, K. Venkatesan, T. Hao, T. Hirozane-Kishikawa, A. Dricot, N. Li, G. F. Berriz, F. D. Gibbons, M. Dreze, N. Ayivi-Guedehoussou, N. Klitgord, C. Simon, M. Boxem, S. Milstein, J. Rosenberg, D. S. Goldberg, L. V. Zhang, S. L. Wong, G. Franklin, S. Li *et al. Nature* **437**, 1173 (2005).

[42] J. Lim, T. Hao, C. Shaw, A. J. Patel, G. Szabo, J. F. Rual, C. J. Fisk, N. Li, A. Smolyar, D. E. Hill, A. L. Barabasi, M. Vidal and H. Y. Zoghbi. *Cell* **125**, 801 (2006).

[43] G. T. Hart, A. K. Ramani and E. M. Marcotte. *Genome Biol.* **7**, 120 (2006).

Chapter Six

Computational Tools to Analyze CAGE — Introduction to PART II

Carsten O. Daub

Omics Science Center, RIKEN Yokohama Institute, Japan
Email: daub@gsc.riken.jp

Recent developments in high-throughput DNA sequencing technology put tag-based protocols like SAGE and CAGE into a new perspective. As mentioned in Chapters 1 and 2, the step from exploratory qualitative sequencing of ESTs and full-length cDNA to now quantitative sequencing opens the door for a much broader set of applications. The large number of small CAGE tag sequences requires extensive data processing and biological conclusions and findings require statistical methods. Compared to earlier established expression technologies like quantitative RT-PCR and DNA microarrays, CAGE data processing, data formats, and analysis strategies are not yet standardized and development is ongoing in the scientific community. Moreover, CAGE elucidates aspects of the transcriptome previously inaccessible on a genome-wide and unbiased scale, opening avenues to find novel global mechanisms of transcriptional regulation.

The CAGE experimental protocol was designed to capture messages with 5′ CAP sites (Chapter 2). In contrast to probe-based technologies like cDNA or oligonucleotide microarrays, where detectable target sequences are chosen *a priori*, the absence of any presumptions constitutes one of the strong advantages of the CAGE technology: the sequencing of a CAGE library is an unbiased sampling from the capped mRNA population existing in the biological sample. While the theoretical upper limit of

Cap Analysis Gene Expression (CAGE): The Science of Decoding Gene Transcription **edited by P Carninci**
Copyright © 2010 by Pan Stanford Publishing Pte Ltd
www.panstanford.com
978-981-4241-34-2

detectable messages is restricted by the amount of initial mRNA used for the preparation of a CAGE library, the number of CAGE tags sequenced from a library determines the practical limitations.

Among the 2nd-generation sequencing machines, the Titanium Roche 454 sequencer produces long reads close to 500 base pairs. The Applied Biosystems SOLiD and Illumina Solexa sequencer machines, on the other hand, aim at shorter reads of around 50 base pairs with higher throughput (Chapter 5 for details). The early availability of the 454 sequencer has led to its initial application for the development of the deepCAGE (CAGE with *deep* next generation sequencing) technology. However, since the shorter reads of the SOLiD and the Solexa machines fit well with the length of CAGE tags, they are likely to become the main sequencing technology utilized for deepCAGE.

The sequencing of a CAGE library is the entry point to the computational analysis. CAGE tags are obtained using the 5′ end sequences of capped masseges 21 to 27 nucleotides long, depending on the restriction enzyme employed (Chapters 2 and 3), and the primary raw sequence reads obtained from the sequencing machine require processing to separate PCR and linker sequences from the actual CAGE tag. Aligning similar tag sequences to each other enables the estimation of errors for library and sequencing quality control. Chapter 7 "Extraction of CAGE tags and basic quality control" discusses data preprocessing-related issues.

The annotation of the CAGE sequence tags to known transcripts builds the basis for all subsequent data analysis. The alignment of a tag to the genomic location it is derived from determines its transcriptional origin. Several criteria need to be considered for the mapping such as the influence of DNA amplification and sequencing errors. Also, tags mapping to more than one genomic location, multimapping tags, can be assigned to such loci with different weights by considering the transcriptional context. The result of the mapping annotates each CAGE tag to the genomic locus it originates from. Additional annotations like gene models and transcript evidence (EST, full-length cDNA) help to set the CAGE data in its genomic context, as described in Chapter 8 "Setting CAGE tags in a genomic context".

The abundance of CAGE tags at a given genomic location corresponds to the level of expression of the transcripts. This way, CAGE data can be employed as a quantitative measure of gene expression. For microarrays, expression corresponds to the amount of target sequence hybridizing to a predefined probe, either an oligonucleotide probe or a whole cDNA. The selection of the probes of a microarray determines the detectable transcripts. CAGE technology, on the other hand, has a much greater potential by enabling exploration of the transcriptome in an unbiased way. As a consequence, the mapped CAGE tags resemble the total transcriptional activity of the biological sample used. Expression by CAGE is obtained specifically for individual transcription start sites (TSS). The TSSs typically fall into groups or tag clusters (TC) indicating regions of transcriptional activity (CAGE defined TSS, CTSS)(Chapter 10 and Ref. 1). Tag clusters can be defined in various ways resulting either in many narrow distinct starting sites or in larger regions of transcription initiation. The appropriate choice of parameters for the clustering depends on the biological question. A typical definition aims at a clustering level that corresponds to the biological promoter concept. Depending on the clustering of CTSSs used, the accumulated expression for a cluster represents the expression level of a gene according to the gene model employed. The expression of one gene which generates transcripts starting from different exons of that gene can be distinguished. Chapter 9 "Using CAGE data for quantitative expression" introduces the advantages and pitfalls of the CAGE based expression and compares it to well established microarrays.

The complex structure of CAGE tags and their expression on a genome-wide scale make an intuitive understanding difficult. Here, the genomic context gives the appropriate reference frame: transcriptome data like ESTs and cDNAs, various gene models, the abundant information on known genomic features like comparative genomics, and other publicly available CAGE data visualized in genome browsers allow CAGE data to be placed in the context of known information. Furthermore, quantitative aspects of CAGE data for gene expression analysis require data formats comparable to well established formats for microarrays. Chapter 10 "Databases and visualization" describes how to inspect CAGE data in the most informative ways and discusses data format-related issues.

The pooling of CAGE tags from the very same cell type obtained under varying experimental treatments or developmental time courses allows for the recognition of detailed expressional responses of the cells. Changes not only in expression levels but also shifting TSSs can be observed; the latter indicating potential alterations in regulation on the level of the core promoter.[2]

Due to the unbiased nature of the CAGE technology, improvements in quality of available genomes can be of direct benefit for existing CAGE libraries because these CAGE libraries can be remapped to a more recent genome version. With the advent of individual genomes, CAGE tags can be mapped to not only reference genomes but also to those individual genomes separately. Moreover, novel unified representations of populations of genomes of the very same species are under development that account for the variability among the individuals of a species. Here, the development of mapping strategies poses a challenging task.

The sequencing of a CAGE library can be compared to the sampling from the population of capped transcripts. In the light of the increase in sequencing depth from continuously improving sequencing technologies, the detection of lowly expressed transcripts will improve the potential detection of novel promoters of protein coding as well as non-coding genes and other capped molecules. Together with current efforts to reduce the amount of required RNA, the sampling of the RNA population will converge into an exhaustive monitoring of the complete RNA population.

References

[1] P. Carninci, A. Sandelin, B. Lenhard, S. Katayama, K. Shimokawa, J. Ponjavic, C. A. Semple, M. S. Taylor, P. G. Engstrom, M. C. Frith *et al*. Genome-wide analysis of mammalian promoter architecture and evolution. *Nat. Genet.* **38**(6), 626–635 (2006).

[2] H. Kawaji, M. C. Frith, S. Katayama, A. Sandelin, C. Kai, J. Kawai, P. Carninci and Y. Hayashizaki. Dynamic usage of transcription start sites within core promoters. *Genome. Biol.* **7**(12), R118 (2006).

Chapter Seven

Extraction and Quality Control of CAGE Tags

Erik Arner[*] and Timo Lassmann[†]

Omics Science Center, RIKEN Yokohama Institute, Japan
*Email: *arner@gsc.riken.jp, †lassman@gsc.riken.jp*

In this chapter, we explain our experience in handling data after completion of CAGE sequencing reads and extraction of the tags. Procedures related to the extraction of the sequencing tags from the linkers, which depend on the design of the experiment. Linkers can also contain DNA barcodes that can be used to determine the original RNA source in case the samples are mixed. In this chapter we also describe strategies for extraction of CAGE tags from possible contaminating sequences. Additionally, we describe how to assess the quality of the libraries in terms of sequencing quality and to look for potential contamination of sequences that are not derive from RNAs.

7.1 OVERVIEW

CAGE is used to map promoter elements, annotate transcript borders, and explore promoter activities in different biological states.[1,2] One typical application is to compare cells or tissues under normal conditions to those after stimulus or perturbation (Suzuki *et al.*, *Nat. Genet.* **41**(5), 553–562 (2009)). Since CAGE requires PCR amplification, it is desirable to amplify several samples in the same reaction, particularly if the samples are to be compared for expression. As described in Chapter 3, DNA barcodes, that are short nucleotide sequences with arbitrary sequences, are

Cap Analysis Gene Expression (CAGE): The Science of Decoding Gene Transcription **edited by P Carninci**
Copyright © 2010 by Pan Stanford Publishing Pte Ltd
www.panstanford.com
978-981-4241-34-2

ligated to the cDNAs before amplification to separate the CAGE tag sequences corresponding to different initial samples after pooling.

An additional linker set (linkers A and B) is used to sequence the multiple CAGE tags that are ligated in longer concatamers to maximize the sequencing length of the Roche 454 Life Science sequencer. At least 3-4 tags can fit into one such combined read, and this number is growing with the increase in sequencing read length.

Tag extraction is the process of retrieving the actual CAGE tags from the various sequences used during library construction. The process itself is straightforward, but the combination of multiple library sequences and sequence errors can somewhat complicate it.

Here, we define sequence errors broadly as nucleotide mismatches and insertions or deletions (InDels) that cause sequences to be unmatched against the genome, transcriptome or the sequences used in the library construction.

7.2 USING READ QUALITIES AND READ PROPERTIES, PRE- AND POST-EXTRACTION

7.2.1 Background: Types of Sequencing Errors

As a main example for a discussion of sequencing errors we discuss our data on the Roche 454 Life Sciences sequencer (from now called simply "454"). Similar sequencing errors are encountered with other sequencing instruments like the Illumina/Solexa and the SOLiD. The 454 contains internal read quality controls that filter out low quality reads. Reads that pass the internal quality control typically have an average error rate of $0.5 - 1.0\%$, depending on read length. The main cause of sequencing errors in 454 is the presence of homopolymers (i.e. continuous stretches of the same nucleotide), accounting for roughly 40% of all sequencing errors.[3,4]

Correspondingly, the most common types of errors are insertions and deletions (InDels). At the level of individual base calls, InDels constitute approximately 60% of all sequencing errors in the 454. The remaining 40% consists of substitution errors and ambiguous base calls (called "N:s") in roughly equal proportions. For the Illumina/Solexa and SOLiD platforms, error rates are currently between 2% and 5%, and mainly consist of substitution errors.

The presence and type of sequencing errors influence mapping rates, mapping specificity, and computer resources needed for mapping. It is therefore desirable to perform additional quality filtering on the data before entering it into the tag extraction and subsequent mapping pipeline. Additionally, sequence quality features can be used post-extraction for additional quality control.

7.3 PROCEDURES BEFORE TAG EXTRACTION

There are a number of indicators of read quality that can be used for pre-extraction filtering in order to increase the mapping rate and specificity and decrease computer running time. These four indicators are listed below in order of significance.

7.3.1 *Ambiguous Base Calls*

The presence of N's (undetermined bases) in reads is the single most significant predictor of poor overall read quality. The internal 454 quality filter by default discards reads with an N rate over a specified threshold, but independent analysis has shown that even the presence of one N in a read is highly indicative of additional errors. By discarding all reads having one or more N's (usually around 5% of all reads), we can eliminate up to 50% of all errors present in the data set. Illumina/Solexa reads can also contain undetermined bases, which also indicate poor read quality, whereas the SOLiD system filters out such reads internally.

7.3.2 *Unusual Read Lengths*

454 reads with lengths outside the main length distribution, i.e., reads much longer or shorter than expected, also indicate low overall read quality. However, with 454, CAGE tags are concatenated and therefore longer read lengths are expected (CAGE tag length plus linker length), so we must take this into account. If the read length property is to be utilized for filtering, reads should be split up into groups according to how many CAGE tags they are likely to contain. For Illumina/Solexa and SOLiD, read length is not an indicator since all reads in a run are of the same length.

7.3.3 Low Average Read Quality Score

Each base in a 454 read is given a quality score Q (also known as phred score[5]), computed as $Q = -\log10(p(error))$, where p(error) ideally is the estimated probability of an erroneous base call. A score of 20 thus corresponds to an error probability of 0.01, a score of 30 is an error probability of 0.001, and so on. Scores are provided for all bases in a 454 read, and are calculated by estimating error probabilities based on the raw data flowgram signal distribution and noise. The quality scores provided by 454 are in principle monotonous predictors of sequence quality, meaning that a base with a higher quality score has a higher probability of being correct than a base with a lower score. However, scores do not always reflect actual error probabilities, and accordingly they cannot be used as such.[6] Nevertheless, the average score is still a good predictor of read quality, as shown in one study in which reads having an average quality below 25 had noticeably higher error rates than those above 25. Note however that the optimal cutoff depends on read length and homopolymer content, and it therefore may vary between reads and projects. Illumina/Solexa and SOLiD systems also provide similar scoresthat can be used for filtering in the same way as described above.

7.3.4 Presence of Sequencing Errors

The presence of detectable sequencing errors is in itself an indicator of additional sequencing errors. In the CAGE protocol for 454, which involves concatenating tags separated by linkers and optionally bar code sequences, it is possible to count the errors occurring in linkers and bar codes. If there are many such errors this indicates poor overall read quality.

We have found that the most effective predictor is ambiguous base calls: it can remove up to 50% of all errors, is very straightforward to use, and is easy to understand. The other predictors are more difficult to use and may require individual assessment for individual projects. However, for large-scale production of CAGE tags, it is best to consider all of the above predictors when setting up the protocol. Optimal filters can quite easily be determined by studying how mapping rates and specificity vary according to these properties.

7.4 USING QC VALUES AFTER TAG EXTRACTION

As mentioned above, the quality scores provided by the 454 do not correspond to actual error rates. Rather, they mainly indicate homopolymer stretch lengths. Generally, the first base in a homopolymer stretch has a very high quality score, with rapidly decreasing scores for each additional base in the stretch. Thus, each stretch of nucleotides (e.g. an extracted CAGE tag) has a quality profile that directly corresponds to the homopolymer composition of the stretch.

We can exploit this for quality control purposes, specifically, we look for regularly occurring contaminants in the data set. By averaging all quality values at each tag position, we can construct an average quality profile on a regular desktop computer within minutes by simply scanning all tag quality values once, then inspect it manually. Typically, a successful run with no contaminants has a perfectly smooth average quality profile, since it is expected that the incidence of homopolymers at each tag position is fairly random. If one or a few species is present in very high copy numbers, the quality profile will not look smooth and instead resemble the individual quality profile of contaminant tags.

For newer sequencing technologies, quality scores depend much less on the nucleotide sequence. Instead, they tend to mostly depend on cycle number with Illumina/Solexa and on both cycle number and primer number with SOLiD. However, while quality scores may be of limited use for detecting contaminants in these systems, they are still very informative in an assessment of the average quality profile — systematic deviance from normal runs usually indicates problems with instrument settings, reagent quality, or library construction.

7.5 ORIGIN OF SEQUENCE ERRORS

The most common sources of error is that produced by the sequencing instrument itself. Each sequencing instrument has a tendency to have a certain type of error. As introduced in Chapter 5, 454 introduces insertions and deletions at roughly the same frequency as mismatches. Since these errors are introduced at the last stage of the sequencing process, they can affect all types of sequences present in the raw reads. For example, these include barcode sequences, linkers, and the CAGE tags themselves in a 454 library.

Secondly, the amplification steps used during the library construction introduce some errors even if high-fidelity polymerase is used. In contrast to sequencing errors, PCR errors appear more frequently as mismatches. In 454 CAGE libraries, PCR errors are naturally eliminated from the linker sequences since concatenation of tags requires a completely matching restriction enzyme site.

Finally, the oligonucleotides used for library construction may not be 100% pure and some mismatching sequences are expected to be produced in oligonucleotide synthesis. In conclusion, errors can be introduced at various stages of the CAGE library preparation protocol. Although there are several sources of errors, the actual error rate of each contribution is generally low.

7.6   USING SEQUENCE ERRORS TO ESTIMATE CAGE QUALITY

Sequence errors within the library sequences can be used independently to measure the sequence quality of the sequenced CAGE library. A simple way is to count the number of sequences in the raw reads that exactly match linker and barcode sequences, then with one mismatch, next with two and so forth. We can easily derive the frequency of mismatches and InDels from such counts. Since barcode sequences may include all types of errors, the barcode error frequencies can be regarded as the maximum error rate explained by experimental procedures. Likewise, error frequencies in the linkers give an idea of the quality of the sequencing run itself. Theoretically it is possible to use the estimated error frequencies in the library sequences to distinguish likely sequencing errors from SNPs.

7.7   A SIMPLE CAGE TAG EXTRACTION METHOD

Here we focus on the extraction of several CAGE tags from long 454 reads. The extraction of SOliD and Solexa CAGE tags is comparatively simple, because with their shorter read lengths only one CAGE tag is sequenced per read.

Essentially, tag extraction is a simple string-matching problem. The known library sequences used for CAGE library construction have to be subtracted from the reads, leaving the remaining CAGE tags behind. Due the errors described in the previous sections, fuzzy matching or regular expressions have to be used. Here scripting languages, especially Perl, are suitable because they have powerful support for regular expression and are widely used in bioinformatics.[7]

The simplest strategy is to search for all sequences resembling 5′ linker and 3′ linker sequences. In addition, reads have to be searched with the reverse complement of these sequences as tags can be ligated together in any orientation. The orientation of tags can be reconstructed since linkers at 5′ and 3′ ends of the tags are different, including barcode sequences that are only present at the 5′ end of the tag.

Tags can then be extracted by matching the positions of 5′ sequences with the positions of the closest 3′ linker sequence. The easiest way is to start with the first linker in the read, try to identify the corresponding downstream linker, and proceed until you reach the end of the read. A drawback of such a greedy strategy is that an incorrectly identified linker sequence at the start of the read will influence the extraction of the following tags. Alternatively, you can build a model of the whole expected read and find the best global match in all sequenced reads. Although this is theoretically a much better strategy, in practice there are few differences in the extracted tags because most of the reads are commonly of high quality.

In some cases the separation of CAGE and linker sequences becomes difficult due to either a high error rate or the unfortunate instance when a genuine CAGE tag resembles a linker sequence. In such cases the extraction method may erroneously report tags that contain linker sequences. Such falsely extracted tags can become problematic later on if they map to the genome and thus give rise to artifactual transcriptional start sites.

A simple way to avoid such tags is to add an additional filter to the method outlined above by requiring a minimum and maximum length of extracted tags to correspond to either the 20 or 27 nt protocol used. For example, for a protocol producing 27-nt long CAGE tags, you look for the first linker sequence and the corresponding next linker 23 to 30 nt downstream in the extraction of the first tag in a read.

An almost unavoidable artifact of the tag extraction is the presence of additional bases at the start of tags, or more rarely at the end of the tags. Such bases arise from insertions within the library sequences close to the tag boundaries, particularly those deriving from a template-free activity of reverse transcription at the 3′ ends of the cDNA. The most frequent form is one, or sometimes a few, extra Gs at the beginning (5′ end) of the sequenced tag. After mapping, these bases are recognizable as 5′ or 3′ mismatches to the genome. Protocols to rescue them have been proposed in the supplementary information of the paper describing the CAGE usage to map mammalian promoters.[2]

Finally, a generally applicable quality control step involves counting 10-mers in the raw reads and looking for highly over-represented 10-mers in the extracted tags. For example, a 10-mer present in more than 50% of the raw reads is likely to be a linker sequence. If such a 10-mer is present in CAGE tags it is likely they were extracted incorrectly. In particular, linker dimers are easily identified with this filtering.

The quality control and extraction of tags from raw sequences involves many steps and may appear complicated at first glance, but for many of the reads produced these steps are trivial because globally the sequence quality is high. Furthermore, the majority of wrongly extracted or low quality tags are filtered out at the next important step: the mapping to the reference genome. Finally, emerging sequencing technologies and new protocols will have fewer PCR cycles, higher sequencing quality, and high throughput to alleviate these problems and offer more stringent filtering criteria.

Nevertheless, when adopting new library preparation protocols or new sequencing platforms, we recommend that you carefully check each step in order to rule out unexpected types of error and bias that may arise from novel sequencing technology in novel environments. We have not mentioned the possibility of low performance of sequencing kits/platforms and the difficulty of performing efficient quality control with the package provided by manufacturers.

References

[1] P. Carninci *et al.* The transcriptional landscape of the mammalian genome. *Science* **309**, 1559–1563 (2005).

[2] P. Carninci *et al.* Genome-wide analysis of mammalian promoter architecture and evolution. *Nat. Genet.* (2006).

[3] M. Margulies *et al.* Genome sequencing in microfabricated high-density picolitre reactors. *Nature* **437**, 376–380 (2005).

[4] S. M. Huse *et al.* Accuracy and quality of massively parallel DNA pyrosequencing. *Genome Biol.* **8**(7), R143 (2007).

[5] B. Ewing and P. Green Base-calling of automated sequencer traces using phred. II. Error probabilities. *Genome Res.* **8**(3):186–194 (1998).

[6] W. Brockman *et al.* Quality scores and SNP detection in sequencing-by-synthesis systems. *Genome Res.*, **18**(5), 763–770 (2008).

[7] S. Lincoln. How perl saved the human genome project. *The Perl Journal* 1 (1997).

Chapter Eight

Setting CAGE Tags in a Genomic Context

Geoffrey J. Faulkner[1,*] and Sean M. Grimmond[2]

[1] *Roslin Institute, University of Edinburgh, UK*
[2] *Institute for Molecular Bioscience, University of Queensland, Australia*
*Email: *geoff.faulkner@roslin.edu.ac.uk*

The detection and quantification of transcriptional activity on the genome by CAGE requires reliable and high-throughput tag mapping. Here, we discuss the basic considerations of sequence tag mapping and how these apply to CAGE. The current CAGE mapping pipeline is then described in detail, in particular how it has been updated to utilize iterative global exact matching for each tag rather than heuristic algorithm local alignment. Combined with the addition of a novel method to resolve tags that map to multiple genomic locations, the most recent evolution of the CAGE mapping pipeline is faster, more accurate and provides better coverage than the previous approach, as we demonstrate using a sample CAGE library. Finally, the pipeline is easily customized for other sequence tag technologies, suggesting broad utility in the era of next-generation sequencing.

8.1 MAPPING PIPELINES FOR SEQUENCE TAG TECHNOLOGIES

Mapping to a reference genome is a fundamental step in the analysis of high-throughput sequence tag data. This computationally intensive process typically requires a matching algorithm and several pre- and post-processing scripts integrated to form an automated pipeline. If the matching task is sufficiently large,

Cap Analysis Gene Expression (CAGE): The Science of Decoding Gene Transcription **edited by P Carninci**
Copyright © 2010 by Pan Stanford Publishing Pte Ltd
www.panstanford.com
978-981-4241-34-2

parallelization across a cluster of processors may be required, adding to scripting complexity. Ultimately the goal is to systematically identify the most likely genomic origin or origins of any given sequence tag in order to build a genome-wide profile of sequence tag signal for further analysis.

Although the implementation of a mapping pipeline may take considerable effort, the underlying principles are very basic. First, determine the minimum requirements for a match and which matching algorithm to use. Next, format the sequence tag set, such as clustering identical tags, and then align each tag to a reference genome using the chosen matching algorithm and minimum match requirements. To finish, parse the matching results to determine the "best" match location or locations for each tag and format the results in a way that enables downstream analysis such as expression profiling. Throughout this process the main idea is to map every individual tag as well as, or better than, a computational biologist could achieve by an *ad hoc* approach and to achieve this level of quality in a high-throughput system.

To maximize the accuracy of this system, informed decisions must be made when determining matching thresholds and the choice of matching algorithm. In the first instance, tag length and error rate are the most important factors. Specifically, the minimum tag length and maximum error rate will dictate the rate at which tags will map uniquely to the genome. If matches are allowed to be too short or are allowed to contain too many errors the probability of mapping to multiple genomic locations, and therefore being difficult to interpret, becomes very high. Figure 8.1 illustrates this concept. In effect, tags less than 16 nucleotides in length will on average map to a single genomic location in less than 15% of cases if no errors are allowed, whereas those containing more than twenty-five matched nucleotides will map to a single genomic location in more than 90% of cases.[1] If errors are allowed the probability of mapping to a single genomic location decreases accordingly. This is crucial to specifying what constitutes a match to the genome as low thresholds may greatly increase the proportion of multi-map tags, whereas high thresholds may result in a very low percentage of tags that can be mapped at all. Low thresholds will also produce many more matches and therefore be slower to process than high thresholds. Depending on the number of tags, reference genome size and the available computational resources, this can also affect how a

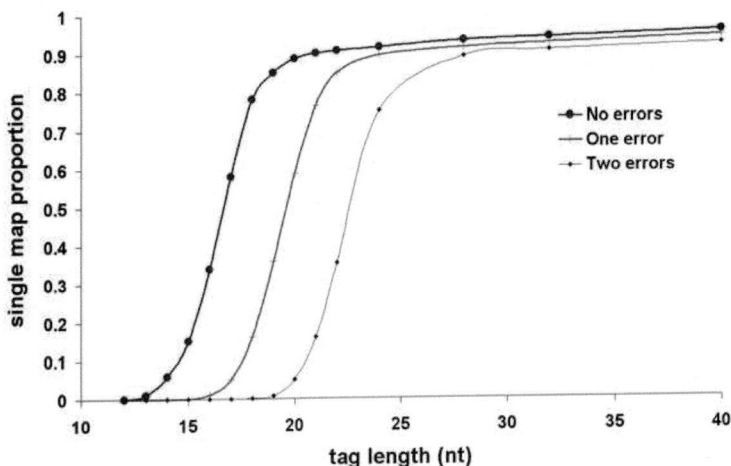

Figure 8.1. Multi-map tag proportion as a function of tag length, when mapped to the human genome and allowing for up two errors over the length of the tag. Tags were extracted from the 5′ end of full-length RefSeq transcripts to simulate CAGE.

match is defined. Matching parameters must therefore reflect a fine balance between coverage, reliability and speed.

Tag length and error rate, as well as total count, also dictate which matching algorithm is used in the pipeline. Heuristic algorithms such as the Basic Local Alignment Search Tool[2] (BLAST) and the BLAST-like Alignment Tool[3] (BLAT) are typically far slower than exact-matching algorithms such as *Vmatch*,[4] are more likely to miss valid matches but can account for more errors per sequence. Consequently, exact-matching algorithms have largely supplanted heuristic approaches in the large-scale mapping of high-quality short sequence tags, such as those generated by the SOLiD and 1G technologies developed by Applied Biosystems and Solexa, respectively.

Once a matching algorithm and associated parameters are decided upon, the mapping pipeline requires scripts to parse and process the information produced by the matching algorithm. The software that accompanies sequencing platforms often integrate these scripts with a proprietary mapping algorithm in what is effectively a "black box," but custom pipelines can achieve similar results using open source matching software and post-processing scripts written in programming languages such Perl, Python or C++. The most important function of these scripts is to identify the "best" match for a tag on the genome, which is usually the alignment containing the highest number of matched bases for that tag, i.e. the longest match with the fewest errors. In the past, tags that matched more than one location at the "best" level (multi-mappers) were at this point removed, though we have shown recently that it is possible to resolve the single most likely location of these tags probabilistically.[1] Finally, matching results are formatted into a table or tables containing sequence tags, match quality, genomic coordinates and other associated annotations for downstream analysis.

8.2 A MAPPING PIPELINE FOR CAGE

The current CAGE mapping pipeline differs substantially from the last published version.[5] In the previous version, tags were matched using BLAST, with the minimum match requirement being an 18 bp alignment with no errors but also with no restriction on errors outside of this 18 bp, regardless of tag length (a local alignment). In the current version, the pipeline is built upon an open source exact-matching algorithm, *Nexalign* (Lassmann *et al.*, in preparation), as well as Python scripts required to format input and output. Nexalign was developed and selected for the pipeline over BLAST due to the comparatively far higher speed of Nexalign and the rapidly expanding CAGE dataset size. In the pipeline, a layered, iterative approach is used to sequentially mutate tags to account for errors and then exact match the modified tags to the genome using *Nexalign*, rather than matching in one pass and specifying error parameters in the alignment algorithm.

To be more specific, identical tags are first clustered into representative tags with an associated count. This dramatically reduces the number of tags to be matched. Secondly, tags are matched exactly to the genome using Nexalign and their positions recorded. As by far the most common error in CAGE tags is the addition of a 5′ guanine during cleavage from cDNA,[5] the next step is to match tags that did not match in the first iteration by excluding their first base and allow for this G-addition error. Those tags that still did not map are subjected to single base pair substitutions (mismatch, insertion, deletion) at every possible position, with each isoform then aligned to the genome using Nex-align. If the tag still does not map the first base is removed to account for a G-addition error and subjected to single base pair substitutions at every position, then aligned to the genome. After this round the alignment that contains the fewest errors for a given tag is designated the "best" match. In this way, the entire length of the tag is considered in a global alignment which includes no more than one error plus one G-addition error.

For the majority of tags the best match is unique on the genome. However, if a tag matches multiple locations at a best match level, a multi-mapping CAGE tag rescue strategy is used to assign it to its most probable location. This method is predicated upon identifying the most likely origin of a tag based upon the unique coincidence of independent tags in close proximity on the genome.[1] Finally, a filter is applied to remove ribosomal RNA-derived tags (where the best match is to an rDNA sequence) prior to clustering and tags-per-million normalization. For an illustrated summary of the pipeline see Fig. 8.2.

8.3 BENCHMARKING WITH A SAMPLE DATASET

To indicate the performance of the current version of the pipeline, a sample dataset, the CAW cloned mouse lung library produced as part of FANTOM3, was mapped to the MM9 version of the mouse genome. 84.5% of the 574,345 tags in the CAW library mapped to at least one genomic position, including 64.4% single map tags, 16.1% rescued multi-map tags and 4.0% multi-map tags that were not rescued (Table 8.1). When the multi-map tags that were not rescued were removed, 80.5% of the library could be reliably mapped to the genome. If likely ribosomal RNA contamination was also removed, this was reduced to 73.3%.

As shown in Table 8.2, we also confirmed that the most common error by far was the addition of an extra guanine to the 5′ end of CAGE tags. The majority of the library (71.7%) matched to the genome with one or fewer errors whilst a further 12.8% matched with two errors (G-addition + one other error). Earlier modeling of the pipeline suggested that only 2-3% more tags would have matched with any two errors allowed, at a disproportionately great computational cost. For this relatively small CAGE library,

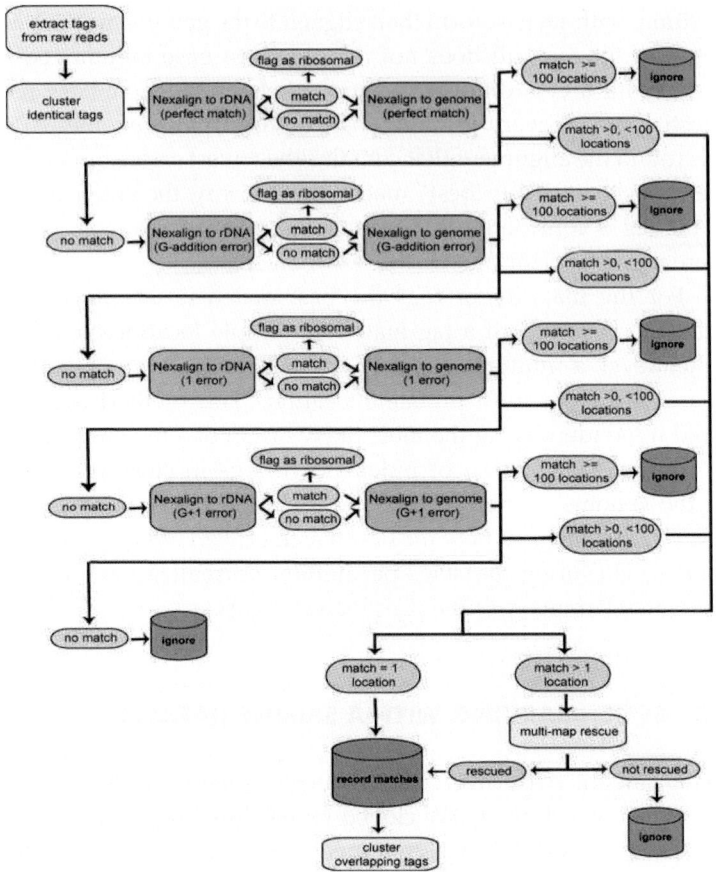

Figure 8.2. Schematic of the CAGE mapping pipeline. Alignment steps are colored in green, decisions in light grey, results in dark grey and processing steps in white.

Table 8.1 Error profiles for mapped tags from the sample CAW FANTOM3 cloned mouse lung CAGE library.

Match Profile	Error	Tag Count	% total
Perfect		100354	20.7
G-addition	error	220523	45.5
any other	error	91060	18.8
G-addition + any other	error	73249	15.0
Total		485186	100

Table 8.2 Mapping results for a sample CAGE library.

Match Type	Tag Count	%total
single	369913	64.4
rescued multi-map	92573	16.1
other multi-map	22700	4.0
unmapped	89159	15.5
total	574345	100

the total processing time was less than 5 processor minutes, including 3 minutes of alignment with Nexalign. By comparison, the previous implementation of the CAGE mapping pipeline would require more than one processor hour to perform the same task.

Moreover, without the multi-map rescue and ribosomal filtering the previous implementation succeeded in mapping only 60.2% of tags reliably to the genome (single-map only), allowing also for less stringent mapping parameters. The relative speed and coverage of the two versions highlights the enhanced utility of the most recent mapping pipeline, for both CAGE and other sequence tag technologies. More importantly, we have demonstrated that it is possible to build a pipeline capable of producing high-quality mappings whilst meeting the increasingly heavy demands of next-generation sequencing.

ACKNOWLEDGMENTS

GJF is supported by an Australian Postgraduate Award through the Australian government Department of Education, Training and Youth Affairs (DETYA). SMG is supported by the ARC Centre in Bioinformatics, the ARC Special Research Centre for Functional and Applied Genomics and an NHMRC Career Development Award.

The CAW library is available from: http: //genomenetwork.nig.ac.jp/public/contents/rel_8th_e.html

References

[1] G. J. Faulkner *et al.* A rescue strategy for multimapping short sequence tags refines surveys of transcriptional activity by CAGE. *Genomics* (2008).

[2] S. F. Altschul, W. Gish, W. Miller, E. W. Myers and D. J. Lipman. Basic local alignment search tool. *J. Mol. Biol.*, **215**, 403–410 (1990).

[3] W. J. Kent. BLAT–the BLAST-like alignment tool. *Genome Res.* **12**, 656–664 (2002).

[4] S. Kurtz, The Vmatch Large Scale Sequence Analysis Software (2006).

[5] P. Carninci *et al.* Genome-wide analysis of mammalian promoter architecture and evolution. *Nat. Genet.* **38**, 626–635 (2006).

Chapter Nine

Using CAGE Data for Quantitative Expression

Michiel De Hoon[1,*], Nicolas Bertin[1] and
Alistair M. Chalk[2]

[1]*Omics Science Center, RIKEN Yokohama Institute, Japan*
[2]*The Eskitis Institute for Cell and Molecular Therapies, Griffith University, Australia*
*Email: *mdehoon@gsc.riken.jp*

Transcriptomics projects utilizing high throughput (HT) gene expression technology provide invaluable landscape information for biological systems. They determine what genes are transcribed, which transcript variants are used and at what expression level. In order to capture as much information as possible about the transcriptome, we advise to use a combination of high throughput methods for determining gene expression. This allows the use of many relatively inexpensive high throughput methods such as microarrays, in combination with discovery based sequencing platforms such as CAGE. The power of these combined datasets allows a greater understanding of the complex nature of the transcriptome, generating many new hypotheses that can subsequently be tested.

9.1 HIGH THROUGHPUT EXPRESSION PLATFORMS

The most common forms of high throughput gene expression used to date are DNA microarray based technologies. These technologies are based on the detection of the specific hybridization of transcripts to complementary oligo-nucleotides fixed onto a substrate, either by differential two color or single fluorescence

Cap Analysis Gene Expression (CAGE): The Science of Decoding Gene Transcription **edited by P Carninci**

Copyright © 2010 by Pan Stanford Publishing Pte Ltd
www.panstanford.com
978-981-4241-34-2

based detection system.[1,2] A wide variety of platforms are now available using different array technologies, from custom made spotted cDNA or PCR products to commercially mass produced oligo-nucleotide sets of various lengths and levels of redundancies spanning from a specific region, generally located in the 3′ end, to the entire coding region of transcripts. Microarray technologies share the same basic principles (for a complete review of available technologies we refer the reader to Kawasaki et al.[3] and the references therein; also of interest is the recent comparison of different microarray platforms by Kuo et al.[4]). These represent very mature technology platforms, with increasing levels of technical replication and coverage.[3–5] However, outstanding issues with microarray technologies remain. In particular, their sensitivity and their detection range dynamics are limited, and their reliance on a known and/or predicted set of transcripts and a pre-defined transcriptome complexity make them generally unfit for uncovering novel variation in expression levels of alternative splicing forms at a given locus. The microarrays recently developed by Affymetrix that are designed to determine exon-level expression for all exons, such as the Affymetrix exon ST array,[6] already enable the discovery and quantification of expression changes of alternative transcripts use in humans.[7,8] However these methods are relatively new and optimal methods for the analysis of such data including normalization and the correct linking of exon combinations with whole transcripts remain unclear. A valuable resource for the comparison of expression profiles is the availability of readily accessible raw data from published microarray studies. Raw data from multiple platforms is available for download and analysis from the Gene Expression Omnibus[9] and ArrayExpress.[10] Two popular 3′ microarray platforms relevant to this chapter are Affymetrix GeneChip® (particularly U133) and the more recent Illumina® beadchip[11] platform.

Quantitative Reverse-Transcribed Polymerase Chain Reaction (qRT-PCR)[12] relies on the monitoring the incorporation of a fluorescent dye during the exponential amplification cycles of a PCR reaction designed with a primer pair specific to a given transcript. Multiplexing and the use of instruments able to measure reactions in plate format have allowed this technique to be used in a medium to relatively high-throughput fashion. qRT-PCR methods alleviate the limitations of microarray based technologies as they provide measures with far greater dynamic range and sensitivity. This approach is tailored to specific genes and allows the measurement of alternative splicing forms by choosing the right combination of primer pairs. However when compared to microarray based technologies, its inherently high cost has rendered this approach restricted to medium throughput experiments. The largest scale expression profiling approaches relying on qRT-PCR that have been reported are limited to a few thousand transcripts.[13] qRT-PCR approaches have also been widely used in an attempt to compare various expression measurement platforms for specific subsets of genes (extensively reviewed in Yauk *et al.*[5])

The ever-decreasing cost of high-throughput sequencing technologies has rendered exploratory gene detection methods amenable to a throughput compatible with a quantitative assessment of transcript expression levels. Several approaches have been developed throughout the years, varying mainly on the preparation of the RNA sample to be captured by sequencing. We refer the reader to Chapters 1,2 and 4 for a more complete review of SAGE, CAGE, MPSS and ditags technologies. Those technologies have the unique advantage that knowledge of transcriptome structure is not required, and only the reference genomic sequence is needed to define the identity of the expressed transcripts.

9.2 COMPARING CAGE TO OTHER MEASURES OF GENE EXPRESSION

CAGE tag sequencing is a very new approach to measure gene expression. It is therefore essential to compare its performance and characteristics to those of other expression measurement methods such as microarray profiling and qRT-PCR experiments. Comparing high-throughput methods for elucidating gene expression is not trivial, involving a number of potentially confounding issues.

9.3 PLATFORM NORMALIZATION

Normalization strategies aim to remove unknown or irrelevant parameters from the raw data. The data are thereby converted to biologically well-defined quantities, which can be compared between experiments and between platforms. While the principle of normalization remains the same, each expression platform has its own set of parameters to be corrected for.

9.3.1 Microarray Normalization

In microarray experiments, data normalization attempts to correct for various sources of variation, such as the amount of RNA used in each sample, or differences in binding affinities between probes.[14,15] In global normalization, we calculate the gene expression level as a fraction of the total mRNA by dividing the gene expression measurement by the sum of gene expression measurements. This fraction can then be compared between experiments, assuming that the total amount of mRNA is equal in each experimental sample. Whereas global normalization assures that the mean expression level is conserved between samples, in quantile normalization[16] the aim is to make the distribution of the measured expression values the same across experiments. This is done by replacing each measured expression value by the average of the expression values across experiments at the same quantile.

For the comparison between CAGE expression data and Illumina® microarray data below, quantile normalization was used for the microarray data. As in general the probes on the microarray have different binding efficiencies, we cannot compare absolute intensities directly to each other. Microarray gene expression values are therefore often expressed as ratios, in which the probe intensity in one condition is compared to the probe intensity in another condition. Usually, the base-2 logarithm of the ratios is reported.

9.3.2 qRT-PCR Normalization

Normalization of qRT-PCR data is done using standard curves, which are generated from standard samples in the same experimental setup.[17] In contrast to microarrays, this normalization results in absolute copy numbers that can be compared directly between genes and between experiments.

9.3.3 CAGE Normalization

In CAGE, we first extract the CAGE tags from the raw sequence reads, and then count how often each tag occurs. The absolute tag counts are not directly comparable between experiments, as they depend on how many raw sequences were read. The tag counts are therefore expressed as "tags per million" (TPM), defined as the expected count for a particular tag if we had extracted one million raw CAGE tags:

$$TPM = 1,000,000 \times \frac{absolute\_tag\_count}{total\_tag\_count}$$

The relative error in the TPM due to sampling noise is approximately

$$relative\_error = \frac{1}{\sqrt{absolute\_tag\_count}}$$

For example, the sampling error for a tag with an absolute count of 16 is approximately 25%. The conversion into TPM is analogous to the global normalization commonly employed for microarray data. However, unlike microarray data, no conversion into expression ratios is required for CAGE expression measurements.

In summary, qRT-PCR gives us an absolute expression level of an RNA, CAGE gives us a relative expression level of a tag (with respect to the other tags), and microarrays gives us the relative expression of an RNA in one condition compared to its expression in another condition. When comparing gene expression measurements between platforms, additional normalization steps may be needed for consistency.

9.4 REPLICATION

Both biological and technical errors contribute to errors between replicates of CAGE expression profiling experiments. Technical replicates make use of the cDNA batch and therefore only assess the technical reproducibility of the experiment. In biological replicates the complete experiment is repeated independently. As biological replicates are affected both by the noise in the expression measurements as well as by noise in the biological system itself, we expect to find a better correlation between technical replicates than between biological replicates.

Microarrays are mature platforms whose reproducibility has been evaluated extensively.[18–22] In these studies, it was found that both differences in microarray platform as well as differences between laboratories contribute to the variability in gene expression measurements. Larkin *et al.*[21] found that gene expression can be measured reliably across microarray platforms if careful attention is paid to experimental details as well as to the gene annotation and the quality of each assay.

The utility of microarrays for explorative transcriptome analysis is highlighted by the advent of the human and rodent Ex1.0ST Affymetrix exon arrays.[6] As microarrays include increasingly high levels of probes (1.4 million probesets on the HuEx1.0ST Affymetrix human exon array) sufficient replication and filtering to reduce the effects of multiple testing is critical to provide meaningful results.

For CAGE and other sequencing-based strategies, we need to consider the effect of the sampling error on the accuracy of expression measurements. At current sequencing depths, tag sequencing has not yet achieved saturation, causing sampling noise in each replicate. At a fixed total amount of sequencing, the introduction of many replicates reduces the number of CAGE tags per replicate and therefore increases the noise in each replicate. Given the cost of sequencing, this means that for most experiments a tradeoff should be made between the number of replicates and the sequencing depth of each replicate to best interrogate the system.

New methods are being developed for comparing expression patterns when the data is discrete (e.g. CAGE or SAGE). These techniques use the distribution of the whole dataset to compare individual gene expression across a dataset.[23] Such techniques rely heavily on replication data to improve the statistical significance, although the required level of replication is not routinely available due to the cost and nature of the technology.

9.5   GENE MODELS AND COMPLEX LOCI

9.5.1   Complex Transcription

Transcription is a complex process involving the transcription, splicing (and most commonly translation) of multiple exons of a gene locus into many alternative isoforms.[24–25] In order to sufficiently capture this complex process, experimental methods must be used that either take the current knowledge of the complexity into account (in the case of microarray technologies), or be independent of knowledge of the potential transcript variants, so long as tags can be mapped to a location on the genome, (e.g. CAGE, SAGE and MPSS technologies). In the case of microarrays, probes must be designed to target all known (and often only predicted) variants of each gene, and it is not always possible to design probes which will specifically determine which of the alternative transcripts are being transcribed. For CAGE and SAGE based approaches, it is also difficult to determine which transcript is transcribed, and validation such at 5' RACE[30] is commonly employed to identify and validate the correct transcript found by CAGE. We refer the reader to Chapter 8 where issues regarding gene model complexity are discussed in detail.

9.5.2 *Gene Expression vs. Transcript Expression*

We wish to determine not only gene-level expression (the overall transcription at a gene locus), but also transcript expression — where the expression level of each alternative transcript is accurately quantified. It is necessary to select a suitable gene model from which to connect the expression values from the various platforms. Widely used gene models include automated methods such as UCSC known genes[26] and Ensembl,[27] and manual curation methods such as RefSeq.[28] Confidence in these transcript sets is high, but sensitivity is typically low (19-28%).[29] Semi-automatic methods such as AceView are more inclusive, reflecting the richer view of the human transcriptome, eluded to by analysis of the ENCODE regions of the human genome where 5 times more transcripts were found compared to RefSeq.[29]

Most technologies measure only parts of the transcript, meaning that it is common that the identification of the correct transcript is ambiguous (Fig. 9.1) This has the effect that it is difficult to compare expression of all transcripts, and we must compare only those transcripts where the mapping of platform expression measures to transcript is unambiguous.

9.6 CONSTRUCTION OF CAGE PROMOTERS AND CALCULATION OF GENE EXPRESSION LEVELS

At each position in the genome, we can calculate the CAGE expression level by counting the number of CAGE tags whose 5'

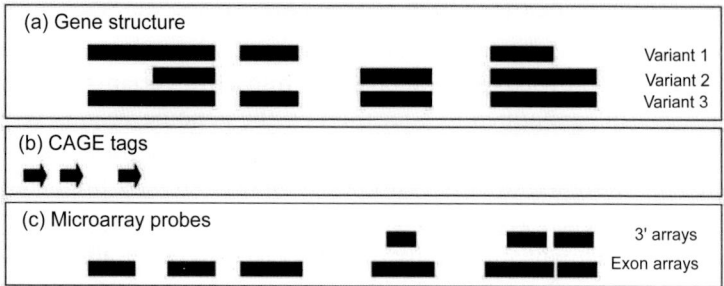

Figure 9.1. The complexity of measuring transcription. (a) An example gene with 3 transcripts variants, showing points of expression measurement by (b) CAGE and (c) microarray platforms. In (b), CAGE can distinguish variant 2, and detects 2 promoters for variants 1 and 3, which in this case are not distinguishable. In (c) 3' microarrays cannot distinguish between variant 2 and 3, but can detect differences between these and variant 1. Exon arrays solve this problem to a large extent, but lack details about promoter usage.

end map to that position. As described above, the expression level is typically expressed as "tags per million". Each genome position with a non-zero tag count can be regarded as a transcription start site. The tag count is not necessarily an integer, since the counts of CAGE tags mapping to multiple locations are divided between the candidate mapping positions. Transcription start sites with a tag count less than one are typically disregarded in the further analysis.

Groups of transcription start sites located close to each other on the same chromosomal strand can be combined together into promoters. Typically, transcription start sites found within 20 base pairs of each other are combined into one promoter. As CAGE promoters often occur in close vicinity of each other, we combine promoters found within a distance of 400 base pairs into promoter regions.

We then assign promoter regions to known transcripts if the promoter region is located close to their 5' end on the same chromosomal strand. Typically, we allow a distance of 500 base pairs upstream of downstream between the promoter region and the 5' end of the known transcript. We then calculate the CAGE-defined expression of genes by summing the tags per million of all CAGE promoter regions assigned to known transcripts that are associated with each gene.

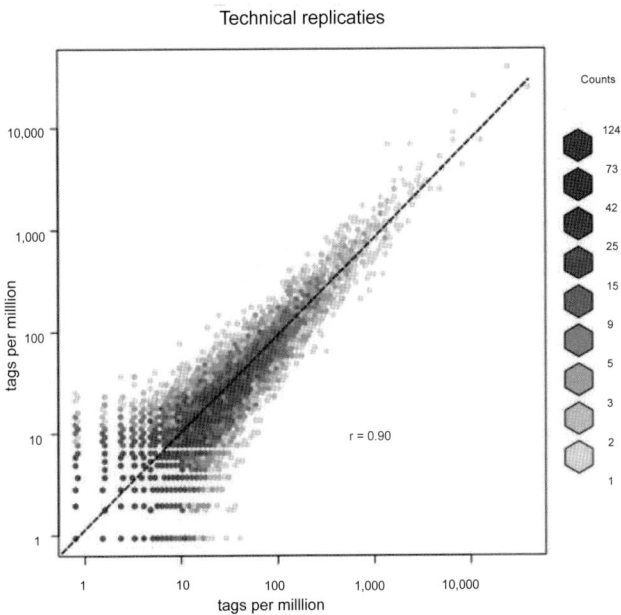

Figure 9.2. Scatter plot of CAGE tags per million per Entrez gene in technical replicates. Each dot represents the number of Entrez genes whose expression value, calculated by summing the counts of CAGE tags mapping within the gene boundaries or in its 1000 bp upstream region, falls into a given bin of expression pair in both compared experiments. The same RNA sample was used for the replicated CAGE experiments.

9.7 COMPARISON OF CAGE EXPRESSION BETWEEN TECHNICAL REPLICATES

To assess the variability introduced by the CAGE protocol, Fig. 9.2 shows, on a logarithmic scale, a scatter plot of technical replicates of a CAGE expression experiment in *Homo sapiens*. Each dot in the scatter plot represents the count of the expression value measured for a single Entrez gene, calculated by summing the CAGE counts of tags generated within the gene boundaries or its 1000 bp upstream region, falling into a given bin of expression values pair. Overall, the technical replicates show a very good agreement, with a Spearman rank correlation equal to 0.90. The agreement between the two replicates remains strong throughout the nearly five orders of magnitude covered in this CAGE expression profiling experiment. At low tag counts, we find a larger variation between the two replicates due to the sampling error. The relative error becomes progressively smaller for genes with higher expression levels.

In the experiment shown in this figure, about 1.9 million tags were sequenced in the first replicate, and approximately 1.6 million tags in the second replicate. About 70% of these tags can be mapped to the genome. For a gene with an expression level of 10 tags per million, this corresponds to 14 tags in the first replicate and 11 in the second replicate, giving a relative error of 25–30% in each replicate due to the sampling error only. For a gene expressed at 10^4 tags per million, the sampling error is less than 1% in each replicate, resulting in a better correlation between the replicates at higher expression levels. The high correlation of 0.90 indicates that the sampling error, being one of the factors affecting the agreement between technical replicates, is acceptably low in this experiment.

9.8 COMPARISON OF CAGE EXPRESSION FROM BIOLOGICAL REPLICATES

When CAGE expression measurements are made using RNA samples from independent biological replicates, we expect to find differences between the CAGE-defined expression levels both due to the CAGE protocol itself and due to true biological variation. Figure 9.3 shows a scatter plot of the expression of Entrez genes as measured in two biological replicates of CAGE expression profiling. The Spearman rank correlation for the scatter plot shown in Figure 9.3 is 0.68. Typically, between biological replicates we find a correlation around 0.60 to 0.65 in the expression of genes as measured by CAGE.

The biological replicates shown on the horizontal axis and the vertical axis in this figure are based on about 1.3 million and 0.8 million mappable CAGE tags, respectively. Accordingly, a gene that is lowly expressed at 10 tags per million has an absolute tag count of 13 and 8 tags, respectively. The corresponding relative error in these tag counts is 28% and 35%, respectively, taking only the sampling error into account. Near 10^3 tags per million, the sampling error is 2.8% and 3.5% in the two replicates.

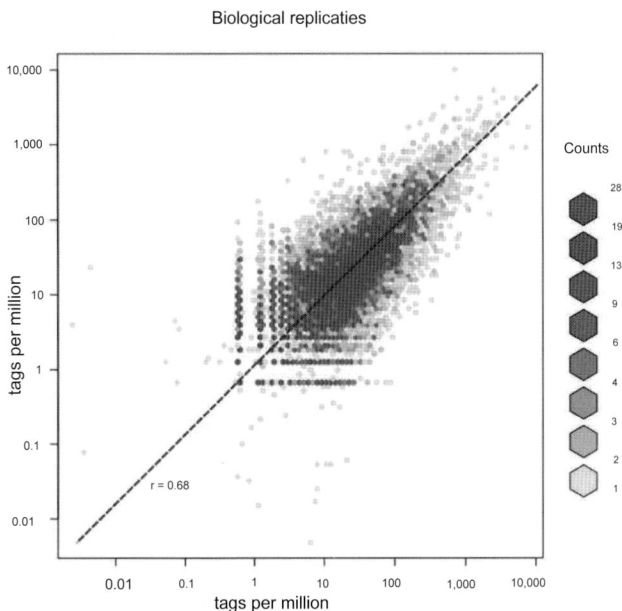

Figure 9.3. Scatter plot of CAGE tags per million per Entrez gene in biological replicates. Each dot represents the number of Entrez genes whose expression value, calculated by summing the counts of CAGE tags mapping within the gene boundaries or in its 1000 bp upstream region, falls into a given bin of expression pair in both compared experiments. The biological replicates assess both the experimental noise as well as the biological noise inherent in the system.

9.9 COMPARISON OF CAGE EXPRESSION BETWEEN DIFFERENT TIME POINTS WITHIN A SINGLE TIME-COURSE

As shown above, technical replicates of CAGE expression profiling show a considerably better agreement than biological replicates. This suggests that biological noise inherent in the system contributes more to the variation in the expression measurements than the technical noise introduced by the CAGE protocol itself.

Often, the biological noise is due to the biological process proceeding at different speeds between replicates. To remove this contribution of biological noise in the comparison, we can consider the agreement in the expression values measured for the same gene at different time points in a time course experiment.

In Fig. 9.4, we show a scatter plot of the expression measured by CAGE for Entrez genes between the first and the last time point in a time course experiment. This figure shows that the expression measurements are well-correlated along the time course, with a Spearman rank correlation of 0.69. This good agreement is partly due to the contribution of genes that are not affected by the experimental manipulation in this time course, for which we expect the expression values at the two time points to coincide. With 1.25 million mappable tags at the first time point, and 460,000 mappable tags at the last time point, we expect a sampling error

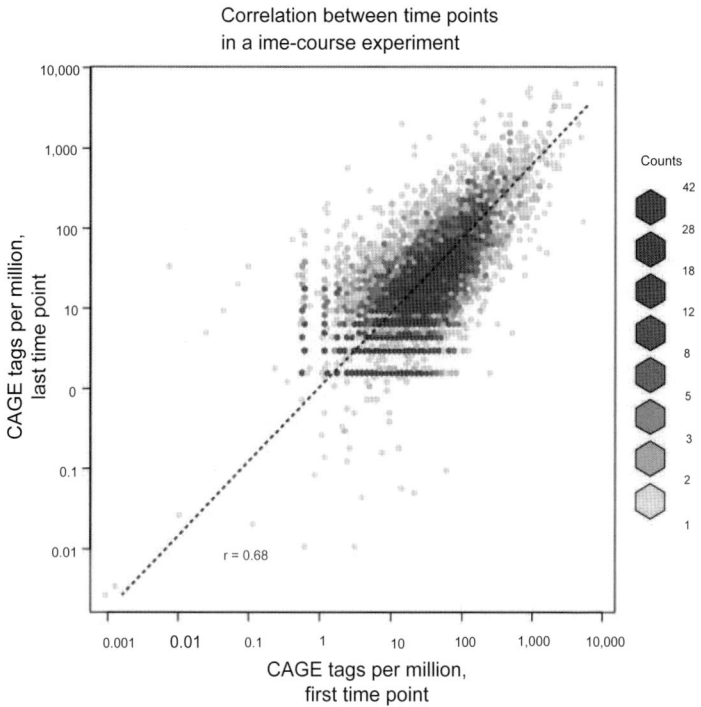

Figure 9.4. Scatter plot of CAGE tags per million per Entrez gene between two time points in a time-course experiment. Each dot represents the number of Entrez genes whose expression value, calculated by summing the counts of CAGE tags mapping within the gene boundaries or in its 1000 bp upstream region, falls into a given bin of expression pair, with the horizontal axis showing the tags per million at the first time point, and the vertical axis the tags per million at the last time point of the experiment.

of 28% and 47% at the two time points respectively for a gene with 10 tags per million, and about 9% and 15% for a gene with 1000 tags per million.

9.10 COMPARISON OF CAGE EXPRESSION PROFILING TO QRT-PCR EXPRESSION MEASUREMENTS

The scatter plot in Fig. 9.5 shows a comparison of the RNA copy number as measured by qRT-PCR, and the CAGE tags per million for the corresponding gene. As qRT-PCR data are absolute expression values, they can be compared directly to the CAGE-derived expression values. In this comparison, we find a Spearman correlation of 0.42. Typical Spearman correlations for CAGE to qRT-PCR comparisons vary between 0.3 and 0.5. Given the 1.25 million mappable CAGE tags in this experiment, we expect a sampling error of 28% at 10 tags per million, and 9% at 1000 tags per million.

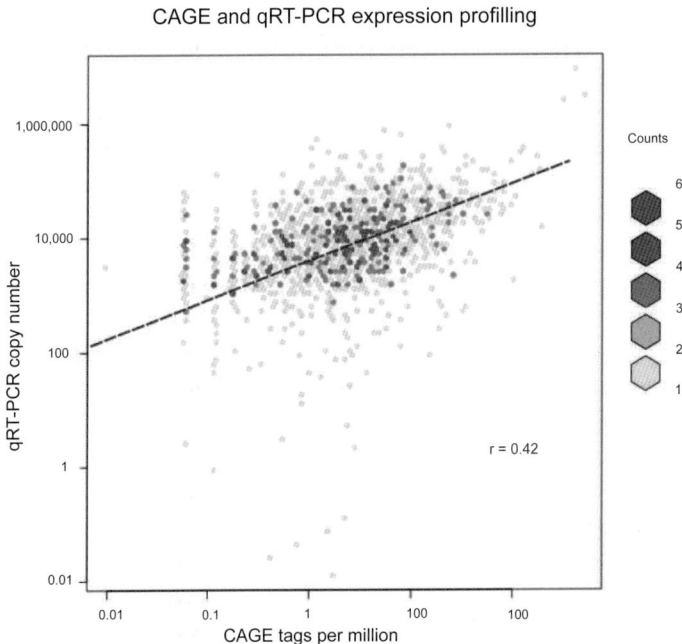

Figure 9.5. Scatter plot of CAGE expression versus qRT-PCR expression measurements. The respective Spearman correlation is 0.42.

9.11 COMPARISON OF CAGE EXPRESSION PROFILING TO MICROARRAY MEASUREMENTS

In order to compare CAGE expression measurements to microarray expression data, for consistency we need to convert the CAGE expression data to expression ratios. Figure 9.6 shows a scatter plot of the expression ratios measured by CAGE and microarrays. For the latter, expression measurements were included only if the detection score was 0.99 or higher. Each expression ratio divides the expression measured for a gene at the last time point of a time course experiment by the expression measured for the same gene at the first time point. As shown in the figure, the expression values measured by CAGE and microarrays agree with a Spearman correlation of 0.69.

This figure is based on the same data as Fig. 9.4. For the expression ratios shown in Fig. 9.6, the sampling errors at the two time points result in a relative error of 55% for a gene with 10 tags per million, and about 17% for a gene with 1000 tags per million.

Whereas CAGE measures the expression level of individual promoters, in microarray experiments the expression levels of the predominantly 3′ probes commonly result in the probes capturing the sum expression of a number of promoters. In particular for genes with many alternative transcripts, it may be difficult to determine which CAGE promoters contribute to which microarray probes. From this viewpoint, we do not expect a perfect correlation between expression measured by CAGE tag sequencing and expression measured by microarray profiling.

Figure 9.6 also demonstrates the larger dynamic range of CAGE expression measurements compared to microarray expression data. Whereas microarray measurements tend to saturate at both ends of the dynamic range, CAGE expression profiling is only affected by the higher sampling error at the low end of the dynamic range.

9.12 PRESENT/ABSENT CALLS

It is common practice to compare ranked lists of expressed genes in different experimental conditions or platforms used. This is a simplification of expression and can be used as an estimate of the concordance between experiments. The main issue with this form

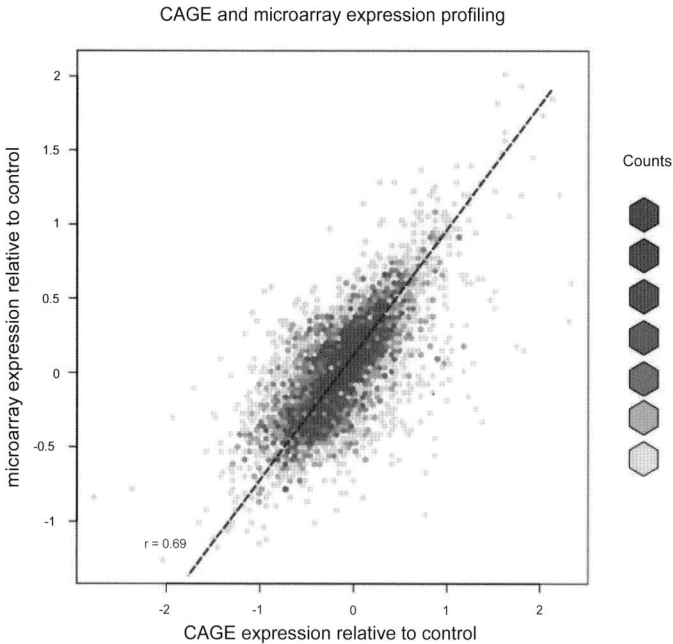

Figure 9.6. Scatter plot of CAGE expression versus microarray expression measurements. For consistency with the microarray expression data, the CAGE expression values were converted to base-2 log-ratios. The Spearman correlation was 0.69.

of comparison is that genes in the "marginal" (or weak) level of expression in either platform make up the majority of the discordance between experiments.

The Venn diagram in Fig. 9.7 shows a comparison of genes detected as present by CAGE and by microarray profiling. A gene is defined to be detected by microarray profiling if its detection score is 0.99 or higher. With this definition, close to 90% of genes detected by microarray profiling are detected by CAGE expression profiling. However, the reverse is not true; almost 40% of genes detected by CAGE expression profiling are not detected in the microarray experiment.

For comparison, Figs. 9.8 and 9.9 show the degree of overlap in detection between two biological replicates for CAGE and microarray profiling, respectively. These figures are based on the same data as those used for Fig. 9.7. The within-platform comparisons show that on average more than 90% of genes detected

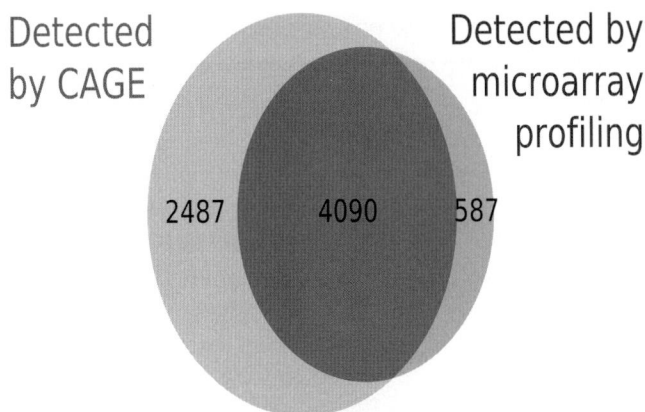

Figure 9.7. Venn diagram showing the concordance between the CAGE and the Illumina® microarray expression profiling platforms. The indicated numbers are the number of genes detected by CAGE or microarray profiling. A gene is detected by CAGE if it has a non-zero expression level; it is detected by microarray profiling if its detection score is equal to or larger than 0.99.

in one of the replicates are also detected in the second replicate. This suggests that the lower concordance in the detection between CAGE and microarray profiling is not due to variability within either platform, but instead reflects that transcripts are assessed differently by the two platforms.

9.13 DISCUSSION

In this chapter we have described the issues that must be considered when using deep CAGE tag profiling for quantitative gene expression. Being a new technology, deep CAGE poses both new challenges and new opportunities for a comprehensive examination of the transcriptome.

Technical and biological replication experiments revealed the amount of variation in expression profiling using deep CAGE experiments. The high correlation between technical replicates suggests that the current sequencing depth is adequate for an accurate quantitation of the gene expression level, though the sampling error can still be sizeable for lowly expressed genes. Some modern sequencers are equipped with multiple sequencing lanes, which can be used directly to assess the sampling error.

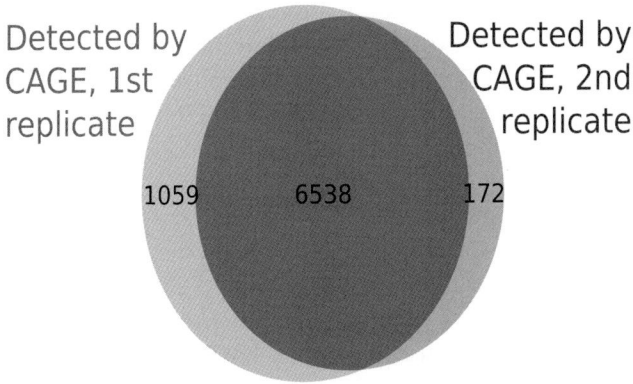

Figure 9.8. Venn diagram showing the concordance of CAGE between two biological replicates. The indicated numbers are the number of genes detected by CAGE expressing profiling. A gene is detected by CAGE if it has a non-zero expression level.

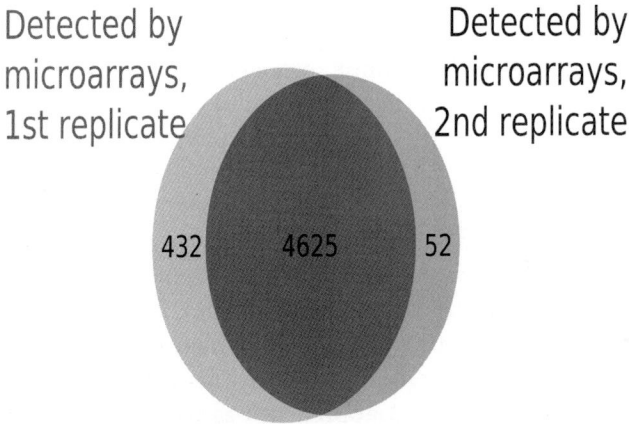

Figure 9.9. Venn diagram showing the concordance of microarray measurements between two biological replicates. The indicated numbers are the number of genes detected by microarray profiling. A gene is detected by microarray profiling if its detection score is equal to or larger than 0.99.

Biological replicates are affected by both the technical variation and the biological variation between experiments, resulting in a lower correlation between two deep CAGE experiments.

To assess deep CAGE as a new technique for expression profiling, it is essential to compare its performance to that of other high-throughput expression platforms. In this chapter, we compared deep CAGE expression profiling to microarray measurements and qRT-PCR experiments. Whereas microarrays have the advantage of being a mature platform, their sensitivity and dynamic range are limited.

Correlations between the three platforms compared (qRT-PCR, microarray and deep CAGE) profiling vary between 0.4 and 0.7. Importantly, measurements made by these platforms need to be interpreted differently. Whereas qRT-PCR produces absolute copy numbers of a particular RNA in the cell, deep CAGE shows the relative contribution of each RNAs to the total cellular RNA content, and microarrays reveal the ratio of the expression of a gene in one condition to its expression in a second condition. To make optimally use of these complementary techniques, the results of different high-throughput profiling methods should be combined in a single comprehensive analysis. This point is underlined by analyzing the detection scores of microarrays and CAGE, showing a larger overlap within platforms than between platforms.

Modern high-throughput sequencers such as SOLiD®, Solexa®, and Helicos® excel at the rapid sequencing of short sequence tags, and are therefore ideally suited for tag sequencing as required for deep CAGE. With the relative sampling error decreasing as the square root of the tag count, modern sequencer technologies allow a better accuracy at a lower cost. We envisage a future scenario where the measurement of gene expression will be based on multiple lines of evidence (such as CAGE, transcriptome sequencing, all exon arrays, 3′ arrays) to interrogate the transcriptome of a cell on a genome-wide scale. This will allow validation between high throughput approaches as well as the generation of hypotheses that can then be validated in smaller-scale experiments on the RNA or protein level. Such a combination of approaches will enable us to properly elucidate the expressed transcripts variants, to compare cell lines and tissues on a deeper level, and to discover the differences in transcription start site usage as well as expressed splice variants.

Mature technologies such as microarray expression have a correspondingly extensive set of analysis tools, of which the bioconductor package is an example.[31] A set of well defined standards for data exchange formats exist for microarray experiments, while standards for sequencing based approaches are still being prepared (see http://www.mged.org). The constant increase in transcriptome sequencing requires the corresponding development of increasingly sophisticated techniques for analysis, data warehousing and visualisation of gene expression data.

Both microarrays and qRT-PCR rely on predefined, known transcripts, and typically have difficulty detecting alternative splice forms, in particular those arising from alternative transcriptional starting sites. Whereas in mouse the available transcriptome structure knowledge is extensive across tissues, this is not the case for human and many other species. Using deep CAGE and other sequencing-based strategies, which are unbiased, one can detect new genes, and novel variants of existing genes and in the case of CAGE variant 5' TSS usage. For loci with a complex transcriptome structure, however, it can be difficult to determine which CAGE tag corresponds to which transcript.

As CAGE is tailored to detect transcriptional initiations, it is the strategy of choice for interrogating the full spectrum of downstream promoter targets of a transcription factor in knock-down experiments. This is in fair contrast with current strategy using microarray platforms where the effect of the knock-down of a transcription factor can only be resolved at the level of genes.

References

[1] D. J. Lockhart *et al.* Expression monitoring by hybridization to high-density oligonucleotide arrays. *Nat. Biotechnol.* **14**(13), 1675–1680 (1996).

[2] M. Schena *et al.* Quantitative monitoring of gene expression patterns with a complementary DNA microarray. *Science* **270**(5235), 467–470 (1995).

[3] E. S. Kawasaki. The end of the microarray Tower of Babel: Will universal standards lead the way? *J. Biomol. Tech.* **17**(3), 200–206 (2006).

[4] W. P. Kuo *et al.* A sequence-oriented comparison of gene expression measurements across different hybridization-based technologies. *Nat. Biotechnol.* **24**(7), 832–840 (2006).

[5] C. L. Yauk and M. L. Berndt. Review of the literature examining the correlation among DNA microarray technologies. *Environ. Mol. Mutagen* **48**(5), 380–394 (2007).

[6] T. A. Clark *et al.* Discovery of tissue-specific exons using comprehensive human exon microarrays. *Genome Biol.* **8**(4), R64 (2007).

[7] P. J. French *et al.* Identification of differentially regulated splice variants and novel exons in glial brain tumors using exon expression arrays. *Cancer Res.* **67**(12), 5635–5642 (2007).

[8] P. J. Gardina *et al.* Alternative splicing and differential gene expression in colon cancer detected by a whole genome exon array. *BMC Genomics* **7**, 325 (2006).

[9] T. Barrett *et al.* NCBI GEO: Mining tens of millions of expression profiles–database and tools update. *Nucleic Acids Res.* **35**(Database issue), D760–765 (2007).

[10] H. Parkinson *et al.* ArrayExpress — a public database of microarray experiments and gene expression profiles. *Nucleic Acids Res.* **35**(Database issue), D747-D750 (2007).

[11] K. Kuhn *et al.* A novel, high-performance random array platform for quantitative gene expression profiling. *Genome Res.* **14**(11), 2347–2356 (2004).

[12] G. Lutfalla and G. Uze. Performing quantitative reverse-transcribed polymerase chain reaction experiments. *Methods Enzymol.* **410**, 386–400 (2006).

[13] Y. Wang *et al.* Large scale real-time PCR validation on gene expression measurements from two commercial long-oligonucleotide microarrays. *BMC Genomics* **7**, 59 (2006).

[14] P. Baldi and G. W Hatfield, *DNA Microarrays and Gene Expression — From Experiments to Data Analysis and Modeling*, Cambridge University Press (2002).

[15] J. Quackenbush. Microarray data normalization and transformation. *Nat. Genet.* **32 Suppl**, 496–501 (2002).

[16] B. M. Bolstad *et al.* A comparison of normalization methods for high density oligonucleotide array data based on variance and bias. *Bioinformatics* **19**(2), 185–193 (2003).

[17] P. Y. Muller *et al.* Processing of gene expression data generated by quantitative real-time RT-PCR. *Biotechniques* **32**(6), 1372–1374, 1376, 1378–1379 (2002).

[18] T. Bammler *et al.* Standardizing global gene expression analysis between laboratories and across platforms. *Nat. Methods* **2**(5), 351–356 (2005).

[19] S. Draghici *et al.* Reliability and reproducibility issues in DNA microarray measurements. *Trends Genet.* **22**(2), 101–109 (2006).

[20] R. A. Irizarry *et al.* Multiple-laboratory comparison of microarray platforms. *Nat. Methods* **2**(5), 345–350 (2005).

[21] J. E. Larkin *et al.* Independence and reproducibility across microarray platforms. *Nat. Methods* **2**(5), 337–344 (2005).

[22] P. K. Tan *et al.* Evaluation of gene expression measurements from commercial microarray platforms. *Nucleic Acids Res.* **31**(19), 5676–5684 (2003).

[23] M. D. Robinson and G. K. Smyth. Moderated statistical tests for assessing differences in tag abundance. *Bioinformatics* **23**(21), 2881–2887 (2007).

[24] E. Birney *et al.* Identification and analysis of functional elements in 1% of the human genome by the ENCODE pilot project. *Nature* **447**(7146), 799–816 (2007).

[25] P. Carninci *et al.* The transcriptional landscape of the mammalian genome. *Science,* **309**(5740), 1559–1563 (2005).

[26] F. Hsu *et al.* The UCSC known genes. *Bioinformatics* **22**(9), 1036–1046 (2006).

[27] T. J. Hubbard *et al.* Ensembl 2007. *Nucleic Acids Res.* **35**(Database issue), D610–617 (2007).

[28] K. D. Pruitt, T. Tatusova and D. R. Maglott, NCBI reference sequences (RefSeq): A curated non-redundant sequence database of genomes, transcripts and proteins. *Nucleic Acids Res.* **35**(Database issue), D61–65 (2007).

[29] D. Thierry-Mieg and J. Thierry-Mieg, AceView: A comprehensive cDNA-supported gene and transcripts annotation. *Genome Biol.,* **7 Suppl 1**, S12 1–14 (2006).

[30] Y. Suzuki and S. Sugano. Construction of a full-length enriched and a 5′-end enriched cDNA library using the oligo-capping method. *Methods Mol. Biol.* **221**, 73–91 (2003).

[31] R. C. Gentleman *et al.* Bioconductor: open software development for computational biology and bioinformatics. *Genome Biol.* **5**(10), R80 (2004).

Chapter Ten

Databases for CAGE Visualization and Analysis

Hideya Kawaji

Omics Science Center, RIKEN Yokohama Institute, Japan
Email: kawaji@gsc.riken.jp

CAGE (Cap Analysis Gene Expression) produces a large number of 5′-end short reads of transcripts, which represent two aspects of transcriptome: transcriptional starting site (TSS) as a part of gene structure and transcriptional initiation activity in a certain biological samples. Computational storage and visualization of them are essential for subsequent analysis and inspection. Here, I discuss databases and visualization tools to handle CAGE data. I review public databases to store such data, and introduce methods enabling to use CAGE data. Genome browser-type database and visualization enable us to analyze transcriptional initiation in relation with other genomic regulatory elements, where transcriptional activity per RNA sample will help examination of context-dependent (or RNA sample specific) transcription. Independent analysis of genomic coordinates provides complementary perspective. Examination of interesting genomic loci and expression patterns using these tools would help us to explore transcriptional events and regulation embedded in the CAGE data.

10.1 INTRODUCTION

The primary output of CAGE (Cap Analysis Gene Expression) protocol[1] is a set of sequences, each of which represents a short reads corresponding to the 5′ end of capped RNA molecules, also called CAGE tags. Subsequent computational processing,

Cap Analysis Gene Expression (CAGE): The Science of Decoding Gene
Transcription **edited by P Carninci**
Copyright © 2010 by Pan Stanford Publishing Pte Ltd
www.panstanford.com
978-981-4241-34-2

such as extraction of CAGE tags from the primary sequences, identification of the genomic location originating the CAGE tags (mapping), their aggregation into a unit of transcriptional initiation on genomic coordinates (clustering) and calculation of its activity or expression level from the number of aggregated CAGE tags, enables us to profile the activity of transcription starting site (TSS) in a given RNA sample. Storage and visualization of the processed results are essential to explore biological events. Here, I introduce the databases for the storing and visualization of the CAGE data. Following the overview of CAGE data in the context of data processing and database, I review databases publicly available and introduce the steps to prepare databases or to visualize of CAGE data.

10.2 TRANSCRIPTION MAPS AND ACTIVITY

CAGE allows analyzing at least two aspects of transcription events in transcriptome level: the genomic locations of transcriptional initiation and the activity of all the core promoters driving specific transcription. The first is obtained from alignment of CAGE tags to the genome, where the genomic coordinate corresponding to the 5′-end of the CAGE tag is the identified TSS. The second, the activity at these transcriptional starting sites, is measured as the frequency of CAGE tags derived from the equivalent transcriptional initiation event. We have to take both aspects into account in the data processing, which differs from other RNA profiling technologies, such as cDNA sequencing or microarray analysis to measure gene expression levels. In full-length cDNA or EST sequencing,[2,3] the number of sequenced molecules has been relatively small due to costs, impairing effective quantitative measurement of expression by measuring their frequency. In fact, these approaches allow mainly deciphering the structures of genes and mere qualitative expression in a given tissue.[4–6] Conversely, with conventional microarrays to measure gene expression, probes are designed to detect constant transcripts regions, and their annotations are usually provided by the microarray supplier, often regardless the possible variants depending on the biological context. As a result, microarrays detect just the difference in the expression between RNA samples. The SAGE (Serial Analysis of Gene Expression)[7] is another method to profile gene expression by sequencing small tags derived from transcripts. As SAGE uses sequencing, it resembles one of the aspects of CAGE. However, the main difference is that internal regions of the transcripts only, usually biased around the 3' ends, can be exclusively achieved with this protocol. Therefore, the major application of SAGE is the detection of expression levels between different RNAs, while the gene structures and regulatory elements cannot be identified.

In contrast, CAGE enables us to profile transcription initiation events with their activities even if they are very specific to the RNA sample. As TSSs vary largely and our knowledge about transcriptional initiation is not yet completed, it is essential to profile TSSs for each RNA sample. A major difference in the data processing, in comparison with the other methods, is the step to identify TSS per assay. This step, including alignment of CAGE tags with the genome sequences, is indispensable because the result is the basis to measure transcriptional initiation activities. It is impossible to define TSSs before the assay, and a database needs to be built on such assay-dependent basis. Due to the nature of mammalian transcription,[8] we think of transcriptional initiation at least at two levels: exact position (base pairs) of transcriptional starts and core promoters. These can be achieved by aggregation of CAGE tags at two levels: CTSS (CAGE defined transcription starting site), which is a cluster of CAGE tags sharing the same nucleotide position at their 5'-ends, and TC (tag cluster), which is a spanning region on the genome representing a set of CTSSs.[8] CTSSs are defined to represent a very fine map of each individual TSS, while the TCs defined a larger area, a unit to identify putative core promoters. As a caveat, caution is necessary because all the TCs may not necessarily represent unique core promoters due to our still limited understanding of fine promoter structures. The number of tags belonging to the cluster constitutes a first measurement of the transcriptional initiation activities, but such count requires normalization to be comparable between RNA samples, because CAGE libraries may be sequenced at different deepness. A simple normalization, a ratio of the tag count belonging to the cluster to the total number of the derived tags from the RNA sample, can be used to compare transcriptional activities between samples, where TPM (tags per million) are used as a unit.[8] A variety of information is associated with CAGE, which require storage and analysis. This include information about the RNA samples (tissue/cell, strain, biological treatments, etc.), derived CAGE sequences, CAGE tags extracted from the sequences, clusters of CAGE tags, tag counts and normalized measurement indicating transcriptional activities in each RNA sample for each CTSS or TC. Among these, the clusters and their normalized tag counts in each RNA sample, corresponding to units of transcriptional initiation and their activities, provide a framework of subsequent analysis, also in connection with the corresponding genes or RNA transcripts.

10.3 PUBLIC DATABASES

Published CAGE tags are available in public databases, where the stored content is different depending on the database (Table 10.1). The primary CAGE data is constituted by a set of nucleotide sequences, and the International Nucleotide Sequence Database Collection (INSDC, http://insdc.org), DDBJ, GenBank

Table 10.1 Public databases storing CAGE and their contents.

Database	CAGE tag sequence	Counts per sample	CAGE tag coordinates	CTSS	TC
INSDC (DDBJ, GenBank, EMBL) MGA division	Yes	Yes	No	No	No
CAGE Basic/Analysis Database	Yes	Yes	Yes	Yes	Yes
Genome Network Platform	Yes	Yes	Yes	Yes	Yes
OmicBrowse	No	No	Yes	No	No
UCSC Genome Browser Database	No	No	No	Yes	No
ENSEMBL	No	No	Yes	No	No

and EMBL, assigns public accession numbers to CAGE tags. Due to the large size of sequence data, the tags are stored in a special division, MGA (Mass Sequence of Genome Annotation: ftp://ftp.ddbj.nig.ac.jp/database/mga/, ftp://ftp.ebi.ac.uk/pub /databases/embl/mga/): CAGE tag sequences and their counts within a specific RNA sample are stored. The data deposited here does not include raw CAGE reads, genomic location identified by the tags and any levels of clusters such as CTSS and TC. Additional sequence analysis is required to understand transcriptional events and the biology behind it. Please note that CAGE data publishing as expression analysis is being discussed later in the section of expression analysis.

At RIKEN, where we have developed the CAGE technology, we developed a set of databases, called CAGE basic and analysis databases,[9] which provide comprehensive information and analysis tools. Three different interfaces are prepared based on the same data set depending on different user requirements: CAGE basic database, CAGE analysis database, and the Genomic Elements Viewer. The first one provides access to the raw results, such as detailed information on RNA samples, computationally extracted CAGE tag sequences, their alignments to the genome, and tag counts. This viewer mainly aims to provide feedbacks to the wet scientists involved in technology development and in monitoring the data preparation. The second one, the CAGE analysis database, aims at providing an analytical tool to biologists. For instance, this viewer enables us to access CAGE tag clusters and their activities in all the RNA samples, to display the various core promoters for given genes and the annotated transcription factor binding sites or other elements in the core promoters. The third one, the Genomic Elements Viewer, is designed to display the CAGE tags in a genomic context, which enables the users to inspect genomic regions with direct access to the above two CAGE databases. This is also essential to connect the CAGE tags to the whole transcriptome in the selected areas. Although these databases are somehow specialized, they also provide useful interfaces to access to comprehensive data produced, which may not be easily retrieved elsewhere.

Only recently, other databases, such as Genome Network Platform (http://genomenetwork.nig.ac.jp/), OmicBrowse,[10] UCSC Genome Browser[11] and ENSEMBL[12] started to provide CAGE data. The Genome Network Platform was launched to store all of the experimental results produced in the Genome Network Project (http://mext-life.jp/genome/), and it stores CAGE as well as other expression profiling. All of the rest are genome browser-type databases, where CAGE is treated as an indicator of transcriptional initiation independently on the information of RNA samples. These databases are advantageous because they contain an integrated view of the identified promoters with other genomic features, such as mRNA, CpG island, genome conservation and a growing number of "omics" data. CAGE compensate for the general lack of markers of transcriptional activity for the various RNA.

10.4 GENOMIC VIEW OF IN-HOUSE DATA

As earlier mentioned, genome browser-based visualization is essential to inspect promoters in association with other genomic features. Although displaying mapping positions of CAGE tags does not allow a comprehensive perspective of expression features, it is essential to incorporate information about the profiles RNA samples so that we can explore biological events. There are two main approaches to visualize CAGE data: (i) to use a public genome browser or (ii) to prepare a dedicated environment. In the first approach, all the required steps are consist simply to convert CAGE data into a suitable format and to upload it in the appropriate genome browser. The second approach allows us further customizing the view and having a specific setting. Here, I introduce the steps to prepare such genomic view for CAGE data, after their mapping onto the genome and clustering into functional groups, which represent the individual TSSs and putative core promoters. In the following examples, all the data, such as CAGE tag clusters and their TPM values, are derived from CAGE analysis database.[9] These cases are outlined below in more details.

The first approach is based on uploading the data into a public database after proper formatting. Here, we use UCSC (University California Santa Cruz) Genome Browser database to visualize clusters based on resolution at single-base positions, or CTSS. These clusters have TPM values to indicate transcriptional initiation activities in the used RNA samples. A histogram, in which x- and y-axes respectively represent genomic coordinates and TPM values, provides a perspective of transcriptional initiation activities on the genome. By presenting such histograms per each analyzed RNA sample, tissue-dependent transcription events can be visualized and inspected (Figure 10.1(a)). BED (Browser Extensible Data) is an available file format for this visualization, where chromosome positions, start, stop, and score are described in order in a tab-delimited format (http://genome.ucsc.edu/FAQ/FAQformat). After the preparation of a file describing genomic coordinates of the clusters and their activity values in the format (Figure 10.1(b)), one can obtain a histogram, like in Figure 10.1(a), by uploading the files via the "custom track". This allows visualizing custom made tracks that cannot be seen by the public but only by the user who uploads the data files. For independent representation of different RNA samples, the data can be split with "track" lines as shown in Figure 10.1(b). This will help the inspection of context-dependent promoter usages, such as the case in which different tissues/RNAs/conditions are compared.

The second approach for the visualization is to prepare a dedicated genome browser, which enables us to customize the interface and representation further. One can store and visualize TC within the Generic Genome Browser.[13] The software is easy to configure and is widely used in WormBase,[14] FlyBase,[15] HapMap,[16] the Genomic Elements Viewer as a part of the RIKEN CAGE databases, and many other sites. For the CAGE Genomic Elements Viewer, we assign expression values (TPM) per RNA sample to TCs in order to show their transcriptional initiation activities, in the same way we do also for the CTSSs. As difference with CTSSs, which are defined as the single base where transcription is observed, the widths of TCs can vary depending on the shape of the promoters. Thus, TPM values of proximal TCs among various tissues and conditions might not be comparable directly because TPM values often depend on the spanning length of the cluster on the genome. In other words, larger clusters tend generally to have larger TPMs. Relative change of transcriptional activity and differences in modality of transcription between RNA samples is of interests and its visualization is very important to formulate further biological hypothesis. The first step to visualize such data is to prepare the data file in a similar way to CTSS, but the TPM values of a TC for distinct RNA samples should be described in one single line (Figure 10.2(b)), in contrast to the case of CTSS. The data file (Figure 10.2(b)) is formatted in GFF (Generic Feature Format; http://www.sanger.ac.uk/Software/formats/GFF/GFF_Spec.sh-tml), which is another tab-delimited format for genome annotation and widely used in many software, including the Generic Genome Browser, treating genomic coordinates. These files carry the information of the genome location (specific for each genome assemblies), strand and frequency of appearance of features like the CAGE tags. The data file will be accessible via the browser with its installation and configuration according to the instruction of the software. A small pop-up window can be set up with the use of "balloon" configuration that is available in the version 1.69 or subsequent versions of the software. This can be used to visualize the change of transcriptional initiation activities (Figure 10.2(a)). The graph is implemented with

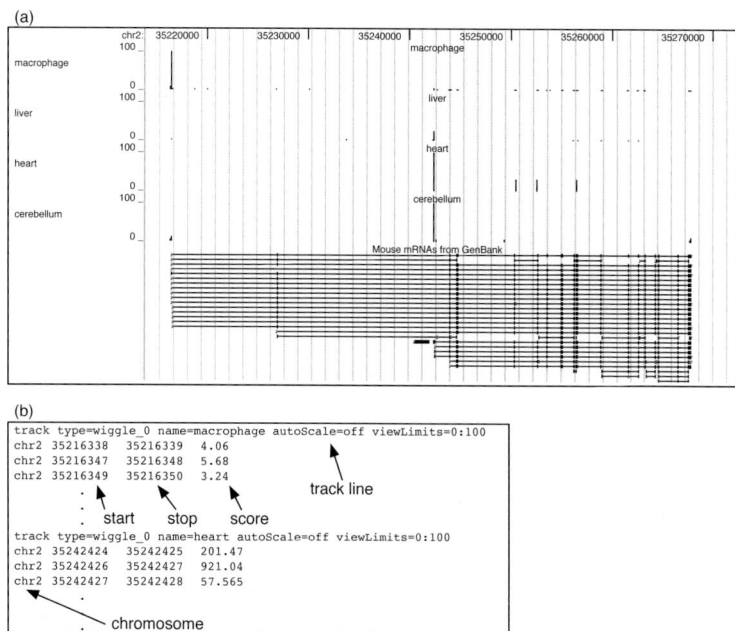

Figure 10.1. Custom track of CTSS on the UCSC genome browser. (a) Histogram representation of CTSS as a custom track of the UCSC genome browser. (b) Data prepared in BED format for the histogram as in (a), where chromosome, start, end, and score are described in a row of tab-delimited format. The data are separated by "track line" to distinguish different tracks (or RNA samples).

additional codes (Figure 10.2(c)) in the configuration file, where Google Chart API (http://code.google.com/apis/chart/) is used. This is just one example of the various possible customizations, and such graphical representations can be set up based on the stored data in the GFF file by adding more pieces of codes to the configuration. The preparation of a given server using the Generic Genome Browser has an additional benefit: the simple configuration enables one to publish the displayed data in DAS

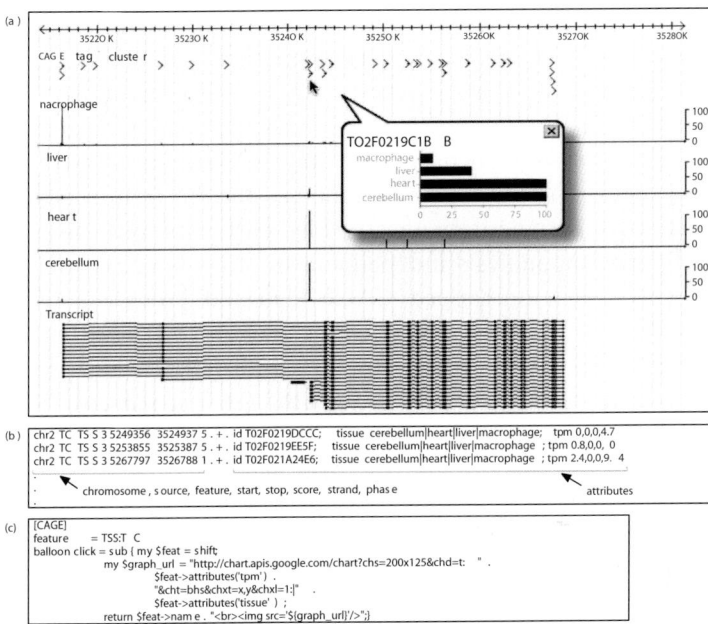

Figure 10.2. Customized representation of TC and CTSS on the generic genome browser (a) Pop-up window to display transcriptional initiation activities in RNA samples with the generic genome browser. (b) Data prepared in GFF format for the interface in (a), where chromosome, source, feature, start, stop, strand, phase, and attributes are described in tab-delimited format. Attributes column in GFF can be used for any types of data, where qualifier and value are joined with space and such pairs are joined with semicolon. In this example, the attributes column store information of ID, profiled tissues, and their TPMs (c) a part of Perl code described in the configuration file, where Google Chart API is called after the tissues and their TPMs, which are described in the attributes column in the data file (b) originally, were obtained from the stored data.

protocol (http://www.biodas.org/) on the Internet as well. The distributed data in DAS can be used by a variety of software used in various browsers, including the ENSEMBL Genome Browser, the Generic Genome Browser and other interfaces.

10.5 FOR EXPRESSION ANALYSES

At difference with inspection at a specific locus, comprehensive analysis of expression similarity and differences between samples cannot be provided with genome browser-type visualizations and independent analysis, such as hierarchical clustering and preparation of heatmaps, are essential. This is also important for a the fraction of CAGE tags that cannot be mapped on the genome for various reasons, such as tags from RNAs produced by duplicated regions on the reference genome, derived from repetitive elements, or derived from unassembled part of the genome.

Expression analysis independent from genomic coordinates can cover such unmapped CAGE tags and provide a whole perspective of expression similarities: this can be performed on a simple data matrix, where rows, columns, and cells represent genes (genomic entities, or unique CAGE tag sequences), RNA samples, and quantified expression value like TPM. A possible problem is the size of data: the data matrix can be too large if one keeps all the unique CAGE tags as row entities. Selection of only fluctuating or interesting entities as usually performed in a microarray analysis would reduce the size of data used in the analysis.

While the biggest value of CAGE is in association with its context of genome sequences, where it is used to retrieve the regulatory elements that are surrounding the CTSSs and TCs (see Chapters 11–17), CAGE provides expression data in analogous ways to microarrays, where the conditions of experiments are very essential. Data release and publication of microarray experiments have to be compliant with MIAME (Minimum Information about a Microarray Experiment) guideline,[17] which has been proposed to ensure the microarray experiments can be universally interpreted and that the derived results can be verified independently. Similarly, while expression profiling based on high-throughput sequencing like CAGE is getting adopted widely due to remarkable advances of sequencing technologies, guidelines and standardization equivalent to MIAME should be adopted in order to release and publish CAGE data in a compliant way. MINSEQ (Minimum Information about a high-throughput SeQuencing Experiment, http://www.mged.org/minseqe/) has been proposed as a guideline for high-throughput sequencing based experiments. We adhere to release CAGE data in a compliant way to this guideline as well as other standards that will appear in the future.

10.6 DISCUSSION

I have discussed two aspects of CAGE, as a marker of transcriptional initiation, hence core promoters, and to monitor the transcriptional activity from a given genomic location for multiple biological conditions. Subsequently, I reviewed public databases that store CAGE data, which can become useful resources for addressing biological questions. The genome view enables us to inspect transcriptional initiation events in association with other elements encoded in the genome, and such view is adopted in many databases. I also introduced the steps to treat any CAGE data, in terms of adapting them to genomic viewers and expression profile. Additional custom tracks on public genome browsers are a first important step, while the preparation of independent servers enables further customization. Data preparation such as the one necessary for genome annotation is a basic step for the both approaches, which enables the user to merge, overlap, and/or combine proprietary data with other public data and genome annotations. Graphical interfaces for such data manipulation have been developed as BioMart (http://www.biomart.org/) and Galaxy (http://g2.bx.psu.edu), and they could be used to explore interesting transcriptional initiation events. A global perspective of expression similarity and differences among tissues, cells and conditions is also an important aspect. Although this can be obtained using conventional expression analysis methods, such as hierarchical clustering and heatmap representation, on a simple data-matrix, I underline the importance to put the expression in the genome context with a global transcriptome and promotome analysis. Reiterative examination of interesting expression patterns and genomic loci would greatly contribute to the experimental and computational biologists interpretation and understanding about transcription as monitored with CAGE.

ACKNOWLEDGEMENTS

This study was supported by Research Grant for the RIKEN Genome Exploration Research Project from the Ministry of Education, Culture, Sports, Science and Technology ofthe Japanese Government to Y.H.; a grant of the Genome Network Project from the Ministry of Education, Culture, Sports, Science and Technology, Japan; Grant for the RIKEN Frontier Research System, Functional RNA research program.

References

[1] T. Shiraki *et al.* Cap analysis gene expression for high-throughput analysis of transcriptional starting point and identification of promoter usage. *Proc. Natl. Acad. Sci. USA* **100**(26), 15776–157781 (2003).

[2] P. Carninci *et al.* Thermostabilization and thermoactivation of thermolabile enzymes by trehalose and its application for the synthesis of full length cDNA. *Proc. Natl. Acad. Sci. USA* A**95**, 520–524 (1998).

[3] K. Maruyama and S. Sugano. Oligo-capping: A simple method to replace the cap structure of eukaryotic mRNAs with oligoribonucleotides. *Gene* **138**(1-2), 171–174 (1994).

[4] T. Hishiki *et al.* BodyMap: A human and mouse gene expression database. *Nucleic Acids Res.* **28**, 136–138 (2000).

[5] J. Kawai *et al.* Functional annotation of a full-length mouse cDNA collection. *Nature* **409**, 685–690 (2001).

[6] Y. Okazaki *et al.* Analysis of the mouse transcriptome based on functional annotation of 60,770 full-length cDNAs. *Nature* **420**(6915), 563–573 (2002).

[7] V. E. Velculescu *et al.* Serial analysis of gene expression. *Science* **270**(5235), 484–487 (1995).

[8] P. Carninci *et al.* Genome-wide analysis of mammalian promoter architecture and evolution. *Nat. Genet.* **38**, 626–635 (2006).

[9] H. Kawaji *et al.* CAGE basic/analysis databases: The CAGE resource for comprehensive promoter analysis, *Nucleic Acids Res.* **34**(Database Issue), D632–636 (2006).

[10] T. Toyoda *et al.* OmicBrowse: A browser of multidimensional omics annotations. *Bioinformatics* **23**(4), 524–526 (2007).

[11] R. M. Kuhn *et al.* The UCSC genome browser database: Update 2007. *Nucleic Acids Res.* **36**, D668–673 (2007).

[12] P. Flicek *et al.* Ensembl 2008. *Nucleic Acids Res.* **36**(Database issue), D707–714 (2008).

[13] L. D. Stein *et al.* The generic genome browser: A building block for a model organism system database. *Genome Res.* **12**(10), 1599–1610 (2002).

[14] A. Rogers *et al.* WormBase 2007. *Nucleic Acids Res.* **36**, D612–617 (2008).

[15] R. J. Wilson, J. L. Goodman and V. B. Strelets. FlyBase: Integration and improvements to query tools. *Nucleic. Acids Res.* **36**(Suppl. 1), D588–593 (2008).

[16] P. C. Sabeti *et al.* Genome-wide detection and characterization of positive selection in human populations. *Nature* **449**, 913–918 (2007).

[17] A. Brazma *et al.* Minimum information about a microarray experiment (MIAME)-toward standards for microarray data. *Nat. Genet.* **29**(4), 365–371 (2001).

Chapter Eleven

Computational Methods to Identify Transcription Factor Binding Sites Using CAGE Information

Vladimir B. Bajic*, Sebastian Schmeier and
Cameron Ross MacPherson
*South African National Bioinformatics Institute (SANBI),
University of the Western Cape, South Africa
Email: *vladimir.bajic@kaust.edu.sa*

CAGE technologies open new avenues to profile transcription start sites (TSSs) of a genome on a large-scale. They synergize the strength of the accurate pinpointing to the location of a TSS, accompanied by the measure of the level of gene expression. These two qualities, when combined with the advanced methods for identifying transcription factors (TFs) and TF binding sites (TFBSs) likely to be involved in gene transcription initiation, pave a way to prioritize links between the predicted TFBSs and the genes in whose promoters they are found. As a consequence, from this data one can derive links of the form TF→TFBS→TSS/promoters→gene and consequently putative transcriptional regulatory networks (TRNs) of genes, TSSs/promoters, TFBSs and TFs affected in the experiment. We present here a methodology that carries out this process and results in the significant reduction of the false positive associations between the predicted TFBSs and TSSs/promoters of the genes, thus resulting in much more accurate TRNs.

Cap Analysis Gene Expression (CAGE): The Science of Decoding Gene Transcription **edited by** P Carninci
Copyright © 2010 by Pan Stanford Publishing Pte Ltd
www.panstanford.com
978-981-4241-34-2

11.1 INTRODUCTION

One of the most useful experimental approaches for deciphering the molecular basis of cellular responses under various conditions is based on gene expression. A differential gene expression can be associated with the different behavior in two cellular conditions, such as expression of normal cells and expression of affected cells both subjected to the same type of conditions. A part of the dynamics of molecular events during a cellular response can also be observed by recording gene expression along a time line, after cells are exposed to a particular stimulus. ch11f01-

There are many experimental technologies that can provide such information, but since this monograph focuses on the use of the cap-analysis of gene expression (CAGE),[10] we will also discuss only methods that are based on CAGE. CAGE tags possess two unique qualities: they synergize the accurate pinpointing to the location of a transcription start site (TSS)[3,4] and the CAGE tag counts also reflect the level of gene expression.[3,4] It is for this reason that the CAGE based experiments to an extent mimic the microarray gene expression experiments, while at the same time CAGE provides additional information about the utilized TSSs within the experimental conditions. Consequently, CAGE experiments can single out a number of genes/transcripts that express in an unusual manner and indicate from which TSSs the transcripts are generated. Expression of these genes is likely the consequence of the conditions of the experiment and thus these genes serve as a guideline for the deciphering of the molecular background of gene transcription during the experiment. Normally, one would consider a differential gene expression to find out the most abnormally expressed genes as potential indicators of the differences considered.

In addition to finding out what set of genes dominantly changes its expression in the experiment, very useful appears to be the information about actually utilized TSSs. Since one mammalian gene loci has normally several promoters and each of the promoters contain several alternative TSSs that are used selectively, for meaningful transcription regulation analysis one need to know not only affected genes, but also the TSSs that are utilized. This information is provided by CAGE and it makes provision for determining the correct set of likely active promoters associated with the implicated TSSs. Moreover, the number of CAGE tag counts found in the experiment can be associated with the level of gene expression: the larger the number of the same tag is found, the more transcripts we expect that contain the tag sequence at its 5'end. For this reason one can treat CAGE tag counts as indicating the level of produced mRNA.

In this chapter we will present a methodology for generating prioritized list of links of the type TF→TFBS →TSS/promoter→ gene, where TF is expected to bind to the TFBS on the gene's promoter determined by the utilized TSS and affect the transcription of the gene in the experiment. This information serves for identifying transcriptional regulatory networks (TRNs). We base this methodology on the CAGE's ability to provide information about the accurate TSS positions, as well as its measure of change of gene transcription initiation events along a time-line. From this we will show how one can prioritize putative links between TFs and the genes they potentially control and use this to enhance predictions of TFBSs and to partly reconstruct TRNs of genes affected in the experiment. The methodology is rather general and results in a significant reduction of false positive associations between TFs and their target genes tch11f01-hrough more accurate association of TFBSs and the genes, thus resulting in much more accurate TRNs.

11.2 SCHEMA OF THE METHODOLOGY PROCESS

To make the presentation of the methodology simpler, we depict in Figs.11.1–11.4 the flow process of the methodological steps required to determine more accurate association of TFBSs and genes, and to reconstruct TRNs. We assume that based on the CAGE data we have determined the set of TSSs associated with the group of genes we are interested in. Once the TSSs are identified, we can determine the associated promoters by extracting sequences upstream and downstream of the TSSs. Our methodology starts from this point.

In the first phase (Fig. 11.1) we map to the target promoters TFBSs of TFs for which we have the necessary computational models. At the same time we map TFBSs also to a set of background sequences. By contrasting the promoter set to the background set we determine the collection of TFBSs that are enriched in the promoter set and determine the parameters (such as the

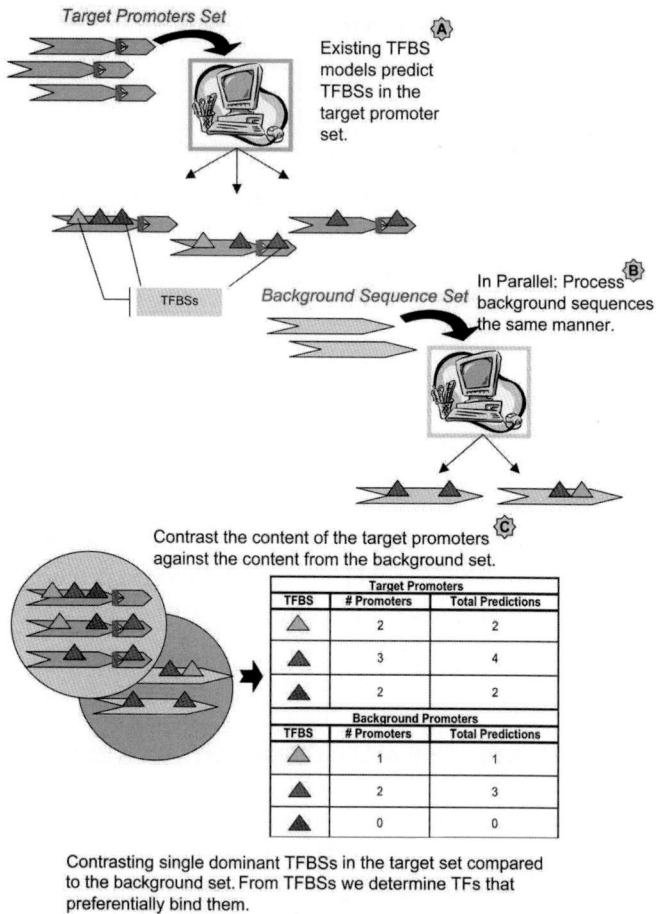

Figure 11.1. Phase 1 of the process.

number of promoters that contain a particular TFBS, total number of predictions of that TFBS, etc.) that characterize each of these TFBS predictions. In the second phase (Fig. 11.2), using the CAGE tag counts along a time line in a cellular response experiment for genes encoding TFs and other genes, we determine potential associations of TFs and their target genes based on the correlation of the respective CAGE tag counts. The strength of that association is given by the absolute value of the correlation

Phase 2: TF->Gene correlation of CAGE tag counts

Determine links between TFs and genes through the correlation of CAGE tag counts over time.

We expect that if a TF controls a gene through its interaction with the gene's promoter, then this is reflected in the correlation of the CAGE tag counts across the time between the gene that encodes for the TF and the affected gene.

The calculated correlation coefficient can be used to rank predicted links of TF ➡ gene.

Note that CAGE tags may be used as an indicator of the degree of gene transcription.

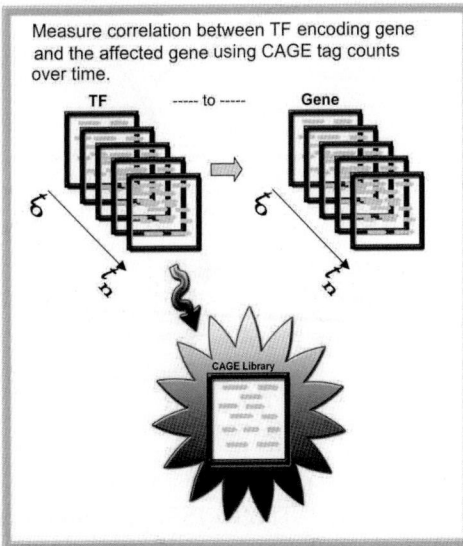

Measure correlation between TF encoding gene and the affected gene using CAGE tag counts over time.

Figure 11.2. Phase 2 of the process.

coefficient (CC). Using correlation information and results of contrasting promoters to the background set (Fig. 11.3), we model our confidence that the TF→TFBS→ TSS/promoter→gene link represent real relationships and we associate to each of these predicted links a number that expresses this confidence. Next we rank all TF→TFBS→TSS/promoter→gene links according to

Phase 3: Refining TFBS predictions

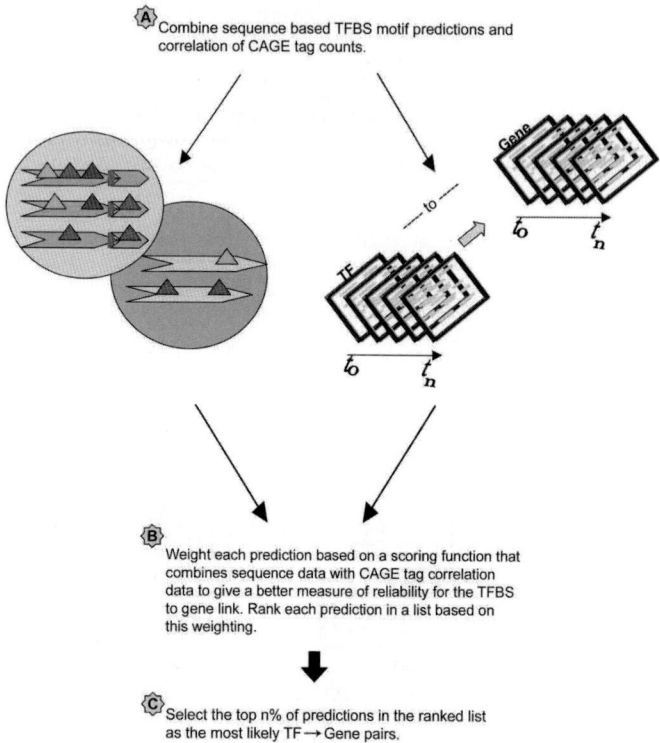

Figure 11.3. Phase 3 of the process.

the confidence and select top n% of these to reconstruct TRNs (Fig. 11.4) using TF→TFBS→TSS/promoter→gene as building blocks.

11.3 INITIAL LINKS OF TF WITH THE AFFECTED GENES

11.3.1 Mapping of TFBSs to Promoters

There is scarce information about the actual interaction of TFs with their target genes via binding of a TF to the regulatory region of the gene and the subsequent modulation of the transcription

Phase 4: Constructing regulatory network

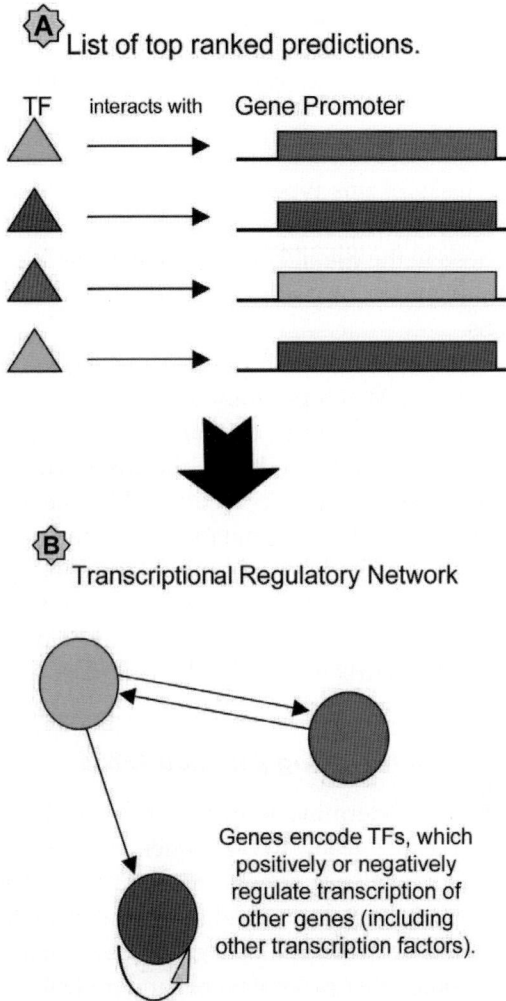

A List of top ranked predictions.

TF   interacts with   Gene Promoter

B Transcriptional Regulatory Network

Genes encode TFs, which
positively or negatively
regulate transcription of
other genes (including
other transcription factors).

Figure 11.4.   Phase 4 of the process.

initiation process as a consequence. One way to putatively infer
such a link is by using predictive models of TFBSs and map
them to the regulatory regions of genes of interest. Two popular
methods are those based on the matching of a consensus sequence
of a TFBS family, and the one based on the use of position

weight matrices (PWMs) that describe TFBS motif families. It is also possible to use other types of TFBS family models, such as Hidden Markov models. In any way, irrespective of the actual tool to be used, TFBSs are predicted on the presumed regulatory region of genes and in this manner such predictions represent potential interaction of TF and TSS/promoter of the gene described by the association model TF→TFBS→TSS/promoter→ gene.

In our experience, the use of PWMs is well suited for large-scale tasks of this type. Several resources are available for this purpose, such as Transfac database[7] and JASPAR.[2] Our preference has been in the use of Transfac database as it is currently the most comprehensive repository of manually curated data on TFs and TFBSs.

The mapping of TFBS models to the genomic sequence can be done by the Match program of the Transfac package.Depending on the goals of analysis, one can use several threshold settings for matching the TFBS models to the DNA sequences. One can use the optimised thresholds as utilized by the Match program provided with the Transfac database.[5] For the more accurate predictions we suggest the use of minFP threshold profiles that aim to ensure minimum proportion of false positive (FP) predictions in the set of predicted TFBSs. These thresholds are determined for each of the matrix models by the Transfac team.

11.3.2 Determining Enriched TFBSs

In order to determine an enriched set of TFBSs we make predictions of TFBSs using PWM models in two groups of sequences: the target sequence set and the background sequence set. For example, the target set could represent promoters of genes we are interested in, while the background set could be composed of a set of some other promoters or random DNA sequences. We will compare the content of the target and background sets in terms of the TFBSs that we mapped to both sets. The idea of comparison is to contrast the two sets in order to single out those TFBSs that are most distinct in one set compared to the other set.[1] We can also associate various numerical indicators of enrichment of TFBSs in target vs. background sets, for example, over-representation index (ORI) as defined in Ref. 1 or p_value of enrichment in the target set relative to the background.

This contrasting process has two important implications. The first one is that it enables us to filter out many predictions of TFBSs that are likely to be the consequence of weak TFBS prediction models and in this way preserve the predictions of a higher accuracy. The other one is that it enables one to determine dominantly present TFBSs in the target set as they are likely to be crucial in driving transcription initiation of the target genes.

After this contrasting procedure we can annotate promoters by using only those TFBSs that have sufficiently high enrichment (high ORI) or that have sufficiently small enrichment p_value. This will provide us the links of the type TFBS→TSS/promoter. We can extend this further to obtain links of the forms TF→TFBS→TSS/promoter→gene.

11.3.3 Score for Confidence of the Predicted TF→TFBS→TSS/Promoter→Gene Association

Although the previously indicated filtering procedure reduces the total number of predictions dramatically, while simultaneously also increases the overall accuracy of the remained predictions, the total number of predicted links of the type TF→TFBS→TSS/promoter→gene is rather high. It is of interest to prioritize the above-mentioned links in some way and one suitable method is to adopt a score measure to express our confidence that TFBS is a real binding site on the promoter of the considered gene. There is no best way to calculate such a score, so we can use some heuristics. The score formula we suggest is:

$$S\_score = (1+log(1/p\_value)) * log(1+ORI) * match\_score$$

Here the *match_score* is a score obtained during matching of the matrix model to the DNA sequence. When we use Transfac matrix models matched by the Match program of the Transfac suite, then

$$match\_score = (corescore + matscore)/2$$

The *corescore* and *matscore* represent the scores effectively obtained by the Match program in matching the PWM model of the most conserved five successive nucleotides of the TFBS motifs and the whole matrix model of the TFBS to the DNA, as explained in Ref. 5

11.4 CORRELATION OF CAGE TAG COUNTS OF GENES AND TFS

Let us assume we have CAGE tag counts obtained in a number of time points $t_1, t_2, \ldots, t_n$, during the experiment. In the normal circumstances we will have more dense time points at the beginning of the monitoring interval, for a simple reason that we want to see in more details what is happening just after stimulus. On the other hand, experiments cost, so we want to economize.

Our interest is in determining the potential association of a TF with the genes whose transcription it may affect. We assume that this effect occurs through the interaction of the TF with the TFBS on the promoter of the affected gene. The association of a TF and a gene can be done, for example, by considering the correlation between the expression of the TF and other genes. Unfortunately, obtaining this information is not straightforward. What we have available is count of CAGE tags at non-uniformly spaced time points. So we need to use this information to find the necessary correlations.

To achieve this we first consider the following. If a TF X controls transcription initiation of gene G this cannot be observed by the associated CAGE tags at any single time point. One need to consider CAGE tags within certain time shift, sufficiently long to allow for TF X to be produced. Roughly, transcript for TF X (the potential existence of which is indicated by the CAGE tag presence) will have to undergo RNA processing, the resulting mRNA has to migrate to cytoplasm and the ribosome needs to translate it into the protein X. This protein then need to migrate back to the nucleus, recognize its relevant binding sites on the DNA control regions of gene G and in combination with other relevant TFs affect the transcription of G. This requires some time to accomplish. The longer is the protein sequence of TF X, the longer time is needed. In practice, 0.5-6 hours would be a reasonable estimate.

For this reason, one should look for the tag count correlations between the gene encoding TF X and gene G within the delay interval of 0.5-6 hours, with G related CAGE tags being delayed.

The first problem is that to determine approximately correct time-delay we need to have counts of tags very frequently and we do not have them. To circumvent this problem we can interpolate the data for the existing tag counts and then resample it uniformly with sufficiently small time steps. Suitable tools for this could be Matlab,[8] Octave,[8] or R.[6] Based on our experience we suggest cubic interpolation as an interpolation method, but other methods could also be used. Once re-sampled tag counts are obtained, we proceed with determining the correlation between X and G. We look for *max(abs(CC))* between the tag counts for X and G allowing for the delay between counts for X and G in the range of 0.5–6 hours, with G being delayed. Here *CC* is the correlation coefficient, while *abs* stands for the absolute value and *max* for the maximum value. Figure 11.5 illustrates the idea of correlation of delayed tag counts.

11.5 RANKING TF→TFBS→TSS/PROMOTER→GENE ASSOCIATION: THE EFFECTIVE USE OF CAGE TAGS

Prediction of TFBSs on the promoter regions of genes of interest contained in the target set opens possibility to infer a potential

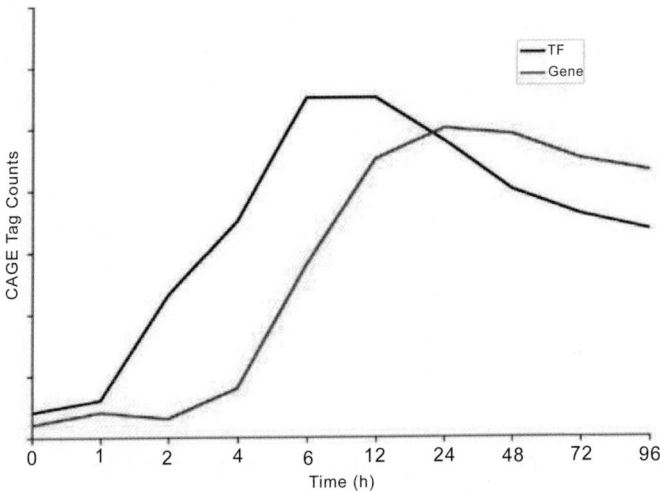

Figure 11.5. Hypothetical change of the tag count distribution by time between X and G and need for considering delayed tag counts for correlation analysis.

of binding of a specific TF to the promoter. However, there are a number of issues of interest to consider in assessing if a TF found in this way has a high chance to represent a real actor in the transcriptional regulation of the considered gene. To find these characteristics we first compare all such found TFBSs in the target promoter set to the background promoter set. This comparison highlights some of the key differences in the observed characteristics for the TFBSs in the target vs. background set. Firstly, if the proportion of the promoters in the target set is higher than in the background set, then the TFBS looks more significant. We associate two parameters with each of these predictions, ORI and enrichment p_values. Both are related to our expectation that the considered TFBS is involved in the regulation of genes in the considered target set. The overall score, S_score, that we associate with the predicted TF$\rightarrow$TFBS$\rightarrow$TSS/promoter$\rightarrow$gene link expresses such expectations in a more formal manner: the higher the score, the more likely we consider it to represent a real relationship. The only connection to the CAGE tags was in the selective determination of the correct TSSs that have generated transcripts.

However, by combining this information with the correlation type based links of the TF$\rightarrow$gene type, we get an enhancement of the sequence based TFBS predictions. The correlation based links TF$\rightarrow$gene are derived under the assumption that the TF will bind to the regulatory region of the gene thus affecting its transcription initiation. Thus, we can combine the S_score and CC to enhance accuracy of the predicted TFBS$\rightarrow$TSS/promoter link obtained purely through the sequence analysis. There are many ways how that can be done, for example, the final score could be

$$Score = S\_score + CC$$

or

$$Score = S\_score * CC$$

Anyway, in this manner the information from the correlation of CAGE tag counts for a gene encoding TF and a gene this TF affects, modulates S_score for the link TF$\rightarrow$TFBS $\rightarrow$TSS/promoter$\rightarrow$gene derived based on matching models of TFBSs to promoters.

The final score $Score$ thus allows for the ranking of TF$\rightarrow$TFBS$\rightarrow$TSS/promoter$\rightarrow$gene links by using information from sequence analysis and CAGE expression analysis. After ranking all predicted links based on *Score*, one can select those above a certain threshold and consider them as the most accurate links predicted in this manner.

This ranking method of TF→ TFBS→ TSS/promoter→ gene (TF→ gene) is new and correlates very well with the statistical significance of the enrichment of TFBSs in the target promoter sets. As a consequence it enables selection of the dominant TFBS→ TSS/promoter (TF→ TFBS→TSS/promoter→gene) links for different target promoter groups. Such filtered best predicted links can be used to build TRNs that will show interactions and information flow of a large number of molecular players in the complex behavior of cell. One should note that, for example, we can select as the target promoter set the full complement of the promoters from one genome and as the background set the random DNA sequences from the same genome and create a unique list of TFBS→TSS/promoter→gene (TF→TFBS→TSS/promoter→gene) associations for the genome. However, looking at smaller promoter groups as the target sets allows for higher resolution of the predicted links

11.6 VERIFICATION OF RESULTS

Any computational analysis requires experimental validation or assessment using existing experimental results. In our case we have been interested in assessing the associations of the type TF→ gene, where TF is controlling transcription initiation of the gene through binding to the respective TFBSs. Our method to arrive to the ranked list of these associations does not guarantee that the implicated links are in fact real. However, to make an assessment of the quality of prediction of the TF→gene association, one could perform the necessary biological experiments to confirm or reject the predicted associations. This, however, is expensive and time consuming. Alternative way is to use data that contains information of the real TF→TFBS→gene associations to serve as a reference for comparison. A good source of such data is the Transfac Professional database http://www.biobase.de/. The other possibility is to compile such data from the published literature. Although this is not fully automatic process, one can get a very good help with the available text-mining tools, such as Dragon

Biomedical Text Miner http://apps.sanbi.ac.za/dbtm/. A user needs to download relevant documents from PubMed and supply them to DBTM. In the DBTM one needs to select two dictionaries "Transcription factors" and "Mammalian genes". The system will then shortlist only those documents that contain explicit information about any of the TFs and genes from the dictionaries contained in DBTM. Moreover, there will be a list of top 5000 associations between any two terms from the two dictionaries. Using these, one needs to manually check (curate) the information from the documents. This process, however, is rather simplified as the relevant documents are readily available with the TFs and genes highlighted.

Once such information is obtained, one can use it to match the ranked list of associations and in this manner get an idea about the quality of the ranked list. In our studies top ranked 500 predicted TF→TFBS→gene links matched ~10%-20% of the experimental information, depending on the type of the target sets.

11.7 RECONSTRUCTION OF TRNS

The top ranked links TF→TFBS→TSS/promoter→gene can be used for reconstruction of the regulatory network. Each link TF→TFBS→TSS/promoter→gene could be split into several links, TF→TFBS, TFBS→TSS/promoter or TFBS→gene. These represent the building blocks of the regulatory network. Alternatively, a link TF→TFBS→TSS/promoter→gene can be contracted to TF→gene if we are not interested to see TFBSs. However, if one is interested in what can be called 'promoter network' the use of TSS/promoter as nodes is necessary, thus the use of links of the type TFBS→TSS/promoter, TF→TSS/promoter, and TSS/promoter→gene, is necessary. Once the building blocks of the regulatory network are determined (TF→TFBS, TFBS→gene, TF→gene, etc), we can use them to build networks that will reflect regulatory relationships between genes and TFs of interest. The process is rather straightforward and the Cytoscape tool[9] is very suitable for visual representation and manipulation of the network layout. The TRNs as described here represent only the partial picture of molecular events in the cell. These networks could be significantly enriched by addition of other information, such as protein-protein interaction.

References

[1] V. B. Bajic, V. Choudhary and C. K.Hock. Content analysis of the core promoter region of human genes. *In Silico Biol.* **4**(2), 109–25 (2004).

[2] J. C. Bryne, E. Valen, M. H. Tang, T. Marstrand, O. Winther, I. da Piedade Krogh A, B. Lenhard and A. Sandelin. JASPAR, the open access database of transcription factor-binding profiles: New content and tools in the 2008 update. *Nucleic Acids Res.* Nov 15 (2007).

[3] P. Carninci, T. Kasukawa, S. Katayama, J. Gough, , M. C. Frith, N. Maeda, R. Oyama, T. Ravasi, B. Lenhard, C. Wells, *et al.* The transcriptional landscape of the mammalian genome. *Science*, **309**, 1559–1563 (2005).

[4] P. Carninci, *et al.* Genome-wide analysis of mammalian promoter architecture and evolution. *Nature Genetics*, **38**, 626–35 (2006).

[5] A. E. Kel, E. Gössling, I. Reuter, E. Cheremushkin, O. V. Kel-Margoulis, and E. Wingender. MATCH: A tool for searching transcription factor binding sites in DNA sequences. *Nucleic Acids Res.* **31**(13), 3576–3579 (2003).

[6] J. Maindonald and W. J. Braun. Data Analysis and Graphics Using R, Second Edition. Cambridge University Press (2007).

[7] V. Matys, O. V. Kel-Margoulis, E. Fricke, I. Liebich, S. Land, A. Barre-Dirrie, I. Reuter, D. Chekmenev, M. Krull, K. Hornischer, N. Voss, P. Stegmaier, B. Lewicki-Potapov, H. Saxel, A. E. Kel and E. Wingender. TRANSFAC and its module TRANSCompel: Transcriptional gene regulation in eukaryotes. *Nucleic Acids Res.* 2006 Jan 1, 34(Database issue):D108–10.

[8] A. Quarteroni and F. Saleri. *Scientific Computing with MATLAB and Octave.* Springer (2006).

[9] P. Shannon, A. Markiel, O. Ozier, *et al* Cytoscape: A software environment for integrated models of biomolecular interaction networks. *Genome Res.* **13**(11), 2498–2504 (2003).

[10] T. Shiraki, S. Kondo, S. Katayama, K. Waki, T. Kasukawa, H. Kawaji, R. Kodzius, A. Watahiki, M. Nakamura, T. Arakawa, *et al.* Cap analysis gene expression for high-throughput analysis of transcriptional starting point and identification of promoter usage. *Proc. Natl Acad. Sci. USA* **100**, 15776–15781 (2003).

Chapter Twelve

Transcription Regulatory Networks Analysis Using CAGE

Jesper Tegnér[1,2,*], Johan Björkegren[1,2,†],
Timothy Ravasi[3-5,‡] and Vladimir B. Bajic[6,§]

[1] *King Gustaf V Research Institute, Karolinska Institutet, Sweden.*
[2] *Department of Physics, Linköping University, Sweden.*
[3] *Scripps NeuroAIDS Preclinical Studies (SNAPS), USA*
[4] *Jacobs School of Engineering, University of California, USA*
[5] *Computational Bioscience Research Centre (CBRC), King Abdullah University of Science and Technology (KAUST), Saudi Arabia*
[6] *South African National Bioinformatics Institute (SANBI), University of Western Cape, South Africa*
*Email: *jesper.tegner@ki.se,[†]johan.bjorkegren@ki.se,*
[‡]*timothy.ravasi@kaust.edu.sa,* [§]*vladimir.bajic@kaust.edu.sa*

Mapping out cellular networks in general and transcriptional networks in particular has proved to be a bottle-neck hampering our understanding of biological processes. Integrative approaches fusing computational and experimental technologies for decoding transcriptional networks at a high level of resolution is therefore of uttermost importance. Yet, this is challenging since the control of gene expression in eukaryotes is a complex multi-level process influenced by several epigenetic factors and the fine interplay between regulatory proteins and the promoter structure governing the combinatorial regulation of gene expression. In this chapter we review how the CAGE data can be integrated with other measurements such as expression, physical interactions and computational prediction of regulatory motifs, which together can provide a genome-wide picture of eukaryotic transcriptional regulatory networks at a new level of resolution.

Cap Analysis Gene Expression (CAGE): The Science of Decoding Gene Transcription **edited by P Carninci**
Copyright © 2010 by Pan Stanford Publishing Pte Ltd
www.panstanford.com
978-981-4241-34-2

12.1 CAGE DATA FOR NETWORK RECONSTRUCTION

Every molecular process occurring in a living cell is a concerted activity of numerous players. Networks, which are defined by the interactions between genes and proteins, govern critical cellular functions such as differentiation and cell death. A great challenge of modern biology is to elucidate these mechanisms and to decipher the corresponding actors in these molecular networks.[20,46] There are, however, numerous layers of interacting molecules and modern experimental and computational technologies have opened new possibilities for obtaining deeper insights into the intrinsic machinery governing the molecular interactions. From a global point of view, one can consider the plethora of interacting molecules as a network of molecular entities that dynamically form under specific intra- and extra-cellular demands and execute their programmed actions accordingly. These networks are characterized not only by the participating molecular entities and their interactions, but also by the direction and dynamics of the interactions.[31,35,47] From this perspective cellular networks are complex systems represented by numerous cause-consequence relationships.

Whole-genome technologies for profiling the molecules within these networks have been instrumental in generating cellular fingerprints obtained during different conditions Monitoring SNPs, mRNA, proteins and metabolites in this manner produces large amounts of data which requires an appropriate computational toolbox for the analysis.[20,38] In this context, the invention of CAGE technology[8,24] has opened yet another chapter into the generation of molecular data that can support biological network reconstruction at a new level of resolution. CAGE tags can also be produced massively under specific cellular and environmental conditions.[7,8,35] In contrast to micro-array gene expression technology monitoring only mRNA levels,[9] CAGE provides unprecedented data on the individual transcription start sites (TSSs) that are effectively utilized during the conditions of the experiment, thus linking the biological process/reaction to the actual control regions of the affected genes. Secondly, CAGE enables a quantitative measure of the gene expression level by utilizing the actual counts of the CAGE tags. In other words, CAGE allows for direct linking the expression events with the effectively utilized regulatory regions of the gene, as well as the actual consequence of the use of these control regions through the measure of transcription/expression via CAGE tags.[7,8]

Clearly, these two advantages position the CAGE technology for providing explicit information necessary for building molecular networks at a new level of resolution as compared to using regular micro-arrays. It is useful in this context to observe a broader schema of the events that lead to the formation and activities of molecular networks.[20] We will present this in a very rough and simplified form. When an external stimulus occurs, sensors on the cell surface react and transmit the chemical information to the interior of the cell. Signal receptors, normally a group of specialized proteins, do activate but could also interact with other down-stream molecules and initiate a set of chain reactions conducting signals to the nucleus, where transcription factors (TFs) and their protein-complexes interact with the DNA initiating transcription of genes. If we pause at this point, we may refer to the entire set of chemical reactions occurring in this process as a signaling pathway and the set of molecular reactions and interactions as a signaling network with the genes being the terminal entities.[46] From the viewpoint of CAGE technology, it provides us data that associates events at the level of interactions between TFs to transcription factor binding sites (TFBSs) on the regulatory regions of affected genes.[35,47] These interactions can be schematically viewed as directed links such as

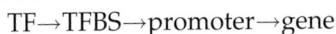

$$TF \rightarrow TFBS \rightarrow promoter \rightarrow gene$$

By considering the gene as a terminal entity in networks we can apply a bottom-up approach for reconstructing the gene part of biologically meaningful molecular networks. Thus, we will not consider the metabolic components in this chapter. In Table 12.1, we put on view the type of entities we will consider for network reconstruction and the corresponding major informatics and genomics resources.

The use of the CAGE tag technology for improving the prediction of TFBSs is discussed in the previous chapter of this monograph. There, it is demonstrated how information provided by CAGE can be combined with the sequence analysis to produce improved predictions of TFBSs, thereby detecting more accurate links of the type.

$$TF \rightarrow TFBS \rightarrow promoter \rightarrow gene.$$

Type of molecules	Source of information
DNA-binding TFs	Genome annotations
	Expression measurements
Cofactors (co-activators/	Genome annotations
repressors; chromatin remodeling)	Expression measurements
Regulatory regions	CAGE TSS mapping
TFBSs	Computational motifs
	search and discovery
	CAGE TSS mapping
Genes	Genome annotations
	Expression measurements

Type of interactions	
TF-DNA (TF-TFBS; TF-	Computational inferences
regulatory region)	Genome-wide location data (GWLA)
PPI (TF-TF; TF-Cofactor)	Protein-protein interaction databases

What is not transparent from CAGE data is how TFs interact with other proteins in the nucleus. It is well known that frequently TFs require other proteins known as co-factors to interact with them and form complexes that are capable of direct binding to DNA. One data-source that can provide such information is represented by protein-protein interactions (PPIs) repositories. By investigating all possible interactions between TFs and other proteins, we obtain a list of putative candidates of proteins that can operate as co-factors. Since the co-factors interact not only with TFs, but also with other proteins, we can gradually expand the network with more distant layers of putative regulatory proteins. Here we aim at the transcription regulatory network (TRN), extended by additional layers of co-factors and complemented by other PPIs.

12.2 METHODOLOGY

To make the presentation of the methodology simpler, we schematically depict it in Fig. 12.1 and describe in detail in the following sections the datasets required.

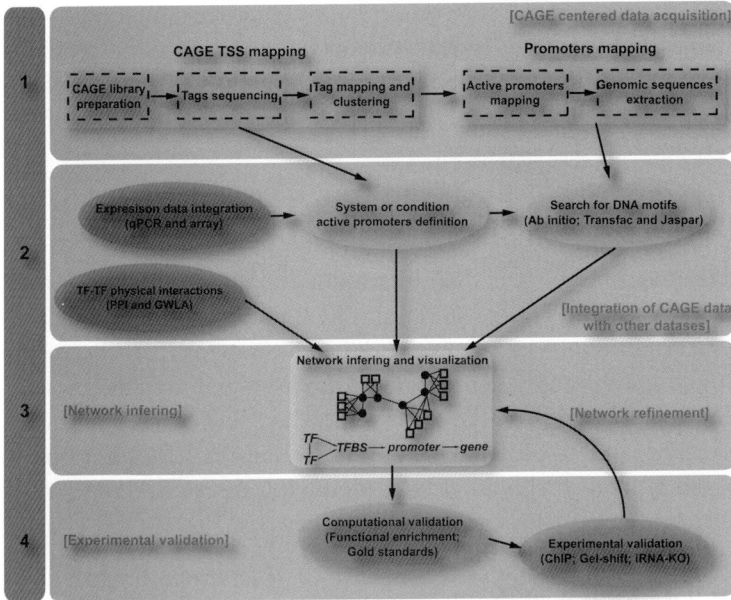

Figure 12.1. Flowchart of the method used to infer CAGE-based transcription regulatory networks (TRN). [Note by the editor: research is in progress to use CAGE data also for expression analysis to determine the expression at each promoter level].

12.2.1 Step 1 of the Process

The first step is the production of CAGE libraries for the system under investigation (see previous chapters). After deep sequencing of these CAGE libraries, the tags are mapped to the genome to determine CAGE defined transcription starting sites (CTSS; Chapter 10). This enables the identification of active starting sites, and hence the promoters for which the genome sequences can be extracted. Furthermore, the number of tags corresponding to the CTSS reflects the expression of the gene associated with vicinity of the CTSS.

12.2.2 Step 2 of the Process

CAGE expression and mapping data are integrated with other expression data in order to infer all TFs and the regulated genes which are expressed by the system, representing the nodes of the network. The promoters of the expressed nodes are then scanned for enriched TFBSs using model based or *de novo* methods. This defines the promoter architecture of the nodes (see the previous Chapter 11). Physical interaction data, such as PPI and protein-DNA interactions, can be also used to increase confidence of the bioinformatics predictions.

12.2.3   *Step 3 of the Process*

The global network is then inferred by combining the nodes using the inferred ($TF{\rightarrow}TFBS$) and physical (PPI and TF-DNA) interactions (edges of the network). The network can be represented graphically where the nodes are usually illustrated by solid shapes and edges are denoted by dashed lines for TF-DNA interaction and as blue lines for PPI. See Fig. 12.2 for an example of the human cerebellum CAGE-derived transcription regulatory network.

12.2.4   *Step 4 of the Process*

The inferred network is then validated using bioinformatics and experimental approaches. The results of validation can be used to refine the original network model.

12.3   GENE EXPRESSION DATA COMPLEMENTARY TO CAGE FOR NETWORK RECONSTRUCTION

Although CAGE provides for digital counting of gene expression, expression profiling using micro-array chips is by far the most popular genome-wide technology for capturing genomic reaction of a cell.[9,41] Expression micro-array is an RNA based method that allows the simultaneous measurement of virtually all the transcripts in a cell. This has been and is still a powerful and wide-spread technique thanks to the relative technical simplicity, low cost, and short turn-over time, which make expression micro-arrays a standard molecular biology technique available to any laboratory. Computational methods used for the analysis of these large collections of data have also been improved and standardized, making the interpretation of micro-array data more accessible to those without a strong computational background.[6,29,38] Although with less throughput than CAGE and chip-based technologies, quantitative real-time PCR

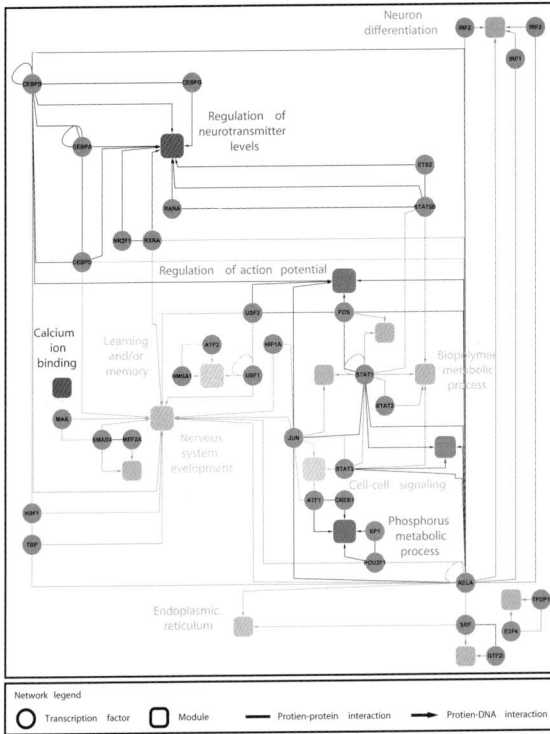

Figure 12.2. A Cerebellar TRN was inferred using CAGE active promoters as described in the text and illustrated in Fig. 12.1 In this particular view, genes expressed in cerebellum are grouped into functional modules and represented as single node (squared nodes) whose size is proportional to the number of genes in the module.

(qRT-PCR) is becoming an increasingly important complementary tool in particular in the construction of TRNs,[35,47] due to its quantitative nature and higher sensitivity which allows for more accurate measurements of low abundant transcripts such as those encoding for transcription factors.[19]

12.4 USING PHYSICAL INTERACTIONS

The edges of a transcriptional regulatory network contain two types of physical interactions: hidden i.e. those between the regulatory proteins and their DNA binding sequences (PDIs) and those between regulatory proteins (PPIs). In eukaryotes, the regulation of gene expression often requires more than one TF to ensure a proper expression of a single gene. TFs interact to form protein complexes and in many cases this is a requirement for the binding of DNA regulatory elements.[4,13,18,23,27,3] For example, this is the case for homo-dimers binding palindromic transcription factor binding sites (TFBS).[49] In the genome, TFBSs tend to cluster together in specific and conserved regions whereas TFs interact at the protein level forming protein complexes that include chromatin remodeling factors.[2,14,30,48] Last, but no less important, are the transcriptional initiation complexes, which despite the fact that they are composed of more than 30 proteins will bind specific regulatory elements via just a few core components such as the TATA box Binding Protein (TBP).[10,25,26,42] The interplay between TFs is often referred to as the combinatorial regulation of gene expression. Therefore capturing all possible combinatorial interactions between TFs is an essential step toward the construction of mammalian transcription regulatory networks. For this purpose the complete maps of PPI are of uttermost value as a first step to map putative pair-wise interactions. PPIs are usually generated by two-hybrid technologies (Y2H).[21] PPI maps can also be constructed using co-immunoprecipitation followed by mass spectrometry.[1,5,12,15,17,28] This technology is more specific than Y2H (less false positive rate) and less scalable. Since the technology relies on co-immunoprecipitation it is more suitable to identify protein complexes with indirect interactions, in contrast to Y2H which instead measures pair-wise, binary and direct interactions.

In recent years the number of binary non-redundant human PPIs has increased dramatically thanks to extensive literature mining (36,617 in the HPRD database)[33,37] and also to large scale experimentally determined human PPIs such as the work from Rual and colleagues and Ewing and colleagues.[12,40] However, one of the limitations with the current human PPI maps is the low coverage of TF interactions because the experimental techniques generally bias toward large macromolecular complexes (i.e. ribosome, splicesosome, membranes channels etc.) and due to a low abundance of TFs compared to cytosolic proteins. Suzuki and colleagues of the RIKEN Genome Science Center in Japan have generated for the first time nuclear specific PPI maps for mouse[44] and now they are focusing on the human nuclear PPI maps (personal communication). Such maps will be very useful resources for the construction of mammalian TRNs. To regulate gene expression, either individual TFs or complexes of TFs need to first bind specifically to cis-regulatory DNA sequences, which are usually at the 5′ end of genes. The most common methods to infer TF-DNA binding events are computational ones (see previous sections).

Technologies have emerged that enable *in vivo* genome-wide experimental mapping of TF-DNA binding events. The most wide-spread of these techniques is the Genome Wide Location Analysis (GWLA) also known as ChIP-chip or ChIP-PET.[11,16,39,51] In GWLA analysis, TF-DNA binding events are captured and frozen in a specific cellular state by *in vivo* crosslinking. Then the genomic DNA is fragmented and the TF of interest is isolated with a specific antibody, along with those genomic fragments bounded by the TF. After crosslinking reversal and protein digestion, the pulled down DNA is labeled in a manner analogous to a cDNA microarray experiment, but hybridized to an oligo micro-array chip whose content is directed towards regulatory regions rather than exons. GWLA are powerful techniques, since they capture *in vivo* and in a high-throughput fashion empirical binding events, thus the TF binding events can be compared across several cellular conditions (drug stimulation, developmental stage, etc.), but contrary to CAGE, they are restricted to one TF at a time.

12.5 TRNS RECONSTRUCTION

The reconstruction of TRN is based on combining the edges as the basic network building blocks. An edge is a link of the form entity1 — entity2, or entity1→entity2. The first type of link is non-directional and is characteristic of PPIs. The second type of link is directional and several types can be derived based on CAGE. As discussed earlier, CAGE can help us elucidate links TF→TFBS→promoter→gene. These can be split into several simpler types, such as TF→TFBS, TF→promoter, TFBS→promoter, TFBS→gene, etc., depending on what details of the interaction we are interested exploring. But we can also expand these by TF→cofactor, TF→protein, cofactor→protein, protein→protein and TF-DNA (ChIP-chip) physical interactions, as mentioned earlier. Using these constituents, blocks of complex TRNs can be reconstructed. Figure 12.2 displays an example of a reconstructed mammalian TRN using data from the CAGE technology (Fig. 12.1).

Clearly the network structure suggests novel mechanistic hypotheses which must be experimentally tested as a final validation step. However, before this step is taken it is mandatory to consider that networks are condition and state-dependent, that is, different parts of the network will be active during different conditions.[31,35] For example, a cell which is exposed to a particular compound or a physiological condition such as stress, will produce two different activity patterns. A network can be generated without using information from expression, but only from 'static' sequence analysis. Such a network we can denote as a static network Therefore, a static network has to be evaluated and projected onto the specific condition of interest. Such a network projection can be performed in space (over different organs/tissues) and/or in time (in response to a stimulation for example) as shown in Fig. 12.2 for human cerebellum.

12.6 USING PATHWAY INFORMATION

Pathways are nothing else but a collection of molecular reactions occurring collectively under specific conditions.[23] In the context of TRNs they are very useful as TRNs can be matched to the pathways and specific segments of TRNs (nodes) could be found in an enriched manner in specific pathways.[34,43] TRN itself will provide a broader context of the pathway functioning and it can suggest possible additional pathway members based on network properties. Vice versa, TRN can be expanded by the other members of the significantly hit pathways: these other not been included in the information from which TRN was reconstructed. On the other hand can pathways provide for the interpretat of the biological role of the TRN and its constitutive elements (Fig. 12.2).

12.7 VALIDATION OF THE RECONSTRUCTED NETWORKS

Since reconstruction of TRNs is a complex process that requires integration and processing of data originating from a variety of resources, it is necessary to make an assessment of the quality of the reconstructed TRNs. At the end, the predictions based on the network analysis have to be evaluated experimentally, but before proceeding to the laboratory there are several useful computational validation steps that can and should be employed. These include statistics, benchmarking against current knowledge, and biological relevance of the annotations associated with the extracted parts lists and pathways.

A proper use of statistics usually goes beyond regular parametric testing including a t-test. As an example, it has become increasingly clear that to evaluate the significance of a particular network motif such as a feed-forward structure, it is necessary to compare the occurrence of the feed-forward loop against a null distribution. Such a statistical randomization procedure can be adapted to the analysis of other features of the reconstructed network, and basically informs us about the significance of a particular finding. Next, it is useful to compare the reconstructed TRN with what is currently known. For example, we can construct a "gold standard list" which we can use for comparison with the predicted TRN. Such a gold standard list is composed of a set of interactions (edges) that have been extensively and experimentally validated and some are available in the literature. A successful validation can be measured using a combined measure of the number of true positive (TP) gold standard interactions recovered (present) in the inferred network as well as those not captured (false negatives, FN) by the predicted TRN. Finally, TRNs can be also computationally validated by biological context relevance, by applying Gene Ontology and Pathways enrichment analyses[3] in order to detect those sub-networks of the TRN that "make sense" in the biology or systems under study. For example, if we are studying a system resembling the brain development, we expect to find regions in the network that are enriched in genes which are involved in processes such as in "nervous systems development", "neuron differentiation", etc.

Although computational validations can be very useful to increase the confidence of the inferred TRNs, it is necessary to assess the novel predicted interactions or regulatory events through functional validation by conducting experiments in the laboratory. Unfortunately, up to date, such an experimental validation is only possible for a handful of targets, because the experiments required to comprehensively assess the biological role and context of a regulatory event are laborious and therefore not yet scalable to a larger number of targets.

References

[1] R. Aebersold and M. Mann. Mass spectrometry-based proteomics, *Nature* **422**, 198 (2003).

[2] J. A. Armstrong and B. M. Emerson. Transcription of chromatin: These are complex times, *Curr. Opin. Genet. Dev.* **8**, 165 (1998).

[3] M. Ashburner, C. A. Ball, J. A. Blake, D. Botstein, H. Butler, J. M. Cherry, A. P. Davis, K. Dolinski, S. S. Dwight, J. T. Eppig, M. A. Harris, D. P. Hill, L. Issel-Tarver, A. Kasarskis, S. Lewis, J. C. Matese, J. E. Richardson, M. Ringwald, G. M. Rubin and G. Sherlock. Gene ontology: Tool for the unification of biology. The Gene Ontology Consortium, *Nat. Genet.* **25**, 25 (2000).

[4] M. Bellorini, D. K. Lee, J. C. Dantonel, K. Zemzoumi, R. G. Roeder, L. Tora and R. Mantovani. CCAAT binding NF-Y-TBP interactions: NF-YB and NF-YC require short domains adjacent to their histone fold motifs for association with TBP basic residues. *Nucleic Acids Res.* **25**, 2174 (1997).

[5] B. Blagoev, I. Kratchmarova, S. E. Ong, M. Nielsen, L. J. Foster and M. Mann. A proteomics strategy to elucidate functional protein-protein interactions applied to EGF signaling. *Nat. Biotechnol.* **21**, 315 (2003).

[6] A. Brazma, P. Hingamp, J. Quackenbush, G. Sherlock, P. Spellman, C. Stoeckert, J. Aach, W. Ansorge, C. A. Ball, H. C. Causton, T. Gaasterland, P. Glenisson, F. C. Holstege, I. F. Kim, V. Markowitz, J. C. Matese, H. Parkinson, A. Robinson, U. Sarkans, S. Schulze-Kremer, J. Stewart, R. Taylor, J. Vilo and M. Vingron. Minimum information about a microarray experiment (MIAME)-toward standards for microarray data. *Nat. Genet.* **29**, 365 (2001).

[7] P. Carninci. Tagging mammalian transcription complexity. *Trends Genet.* **22**, 501 (2006).

[8] P. Carninci, A. Sandelin, B. Lenhard, S. Katayama, K. Shimokawa, J. Ponjavic, C. A. Semple, M. S. Taylor, P. G. Engstrom, M. C. Frith, A. R. Forrest, W. B. Alkema, S. L. Tan, C. Plessy, R. Kodzius, T. Ravasi, T. Kasukawa, S. Fukuda, M. Kanamori-Katayama, Y. Kitazume, H. Kawaji, C. Kai, M. Nakamura, H. Konno, K. Nakano, S. Mottagui-Tabar, P. Arner, A. Chesi, S. Gustincich, F. Persichetti, H. Suzuki, S. M. Grimmond, C. A. Wells, V. Orlando, C. Wahlestedt, E. T. Liu, M. Harbers, J. Kawai, V. B. Bajic, D. A. Hume and Y. Hayashizaki. Genome-wide analysis of mammalian promoter architecture and evolution. *Nat. Genet.* **38**, 626 (2006).

[9] J. DeRisi, L. Penland, P. O. Brown, M. L. Bittner, P. S. Meltzer, M. Ray, Y. Chen, Y. A. Su and J. M. Trent. Use of a cDNA microarray to analyse gene expression patterns in human cancer. *Nat. Genet.* **14**, 457 (1996).

[10] A. Dvir, J. W. Conaway and R. C. Conaway. Mechanism of transcription initiation and promoter escape by RNA polymerase II.
Curr. Opin. Genet. Dev. **11**, 209 (2001).

[11] G. M. Euskirchen, J. S. Rozowsky, C. L. Wei, W. H. Lee, Z. D. Zhang,
S. Hartman, O. Emanuelsson, V. Stolc, S. Weissman, M. B. Gerstein,
Y. Ruan and M. Snyder. Mapping of transcription factor binding
regions in mammalian cells by ChIP: Comparison of array- and
sequencing-based technologies. *Genome Res.* **17**, 898 (2007).

[12] R. M. Ewing, P. Chu, F. Elisma, H. Li, P. Taylor, S. Climie,
L. McBroom-Cerajewski, M. D. Robinson, L. O'Connor, M. Li,
R. Taylor, M. Dharsee, Y. Ho, A. Heilbut, L. Moore, S. Zhang,
O. Ornatsky, Y. V. Bukhman, M. Ethier, Y. Sheng, J. Vasilescu,
M. Abu-Farha, J. P. Lambert, H. S. Duewel, II. Stewart, B. Kuehl,
K. Hogue, K. Colwill, K. Gladwish, B. Muskat, R. Kinach,
S. L. Adams, M. F. Moran, G. B. Morin, T. Topaloglou and D. Figeys.
Large-scale mapping of human protein-protein interactions by
mass spectrometry. *Mol. Syst. Biol.* **3**, 89 (2007).

[13] J. V. Falvo, A. M. Uglialoro, B. M. Brinkman, M. Merika,
B. S. Parekh, E. Y. Tsai, H. C. King, A. D. Morielli, E. G. Peralta,
T. Maniatis, D. Thanos and A. E. Goldfeld. Stimulus-specific assembly of enhancer complexes on the tumor necrosis factor alpha
gene promoter. *Mol. Cell. Biol.* **20**, 2239 (2000).

[14] M. Gaestel. Molecular chaperones in signal transduction. *Handbook
Exp. Pharmacol.* 93 (2006).

[15] A. C. Gingras, R. Aebersold and B. Raught. Advances in protein
complex analysis using mass spectrometry. *J. Physiol.* **563**, 11 (2005).

[16] N. D. Heintzman, R. K. Stuart, G. Hon, Y. Fu, C. W. Ching,
R. D. Hawkins, L. D. Barrera, S. Van Calcar, C. Qu, K. A. Ching,
W. Wang, Z. Weng, R. D. Green, G. E. Crawford and B. Ren. Distinct
and predictive chromatin signatures of transcriptional promoters
and enhancers in the human genome. *Nat. Genet.* **39**, 311 (2007).

[17] Y. Ho, A. Gruhler, A. Heilbut, G. D. Bader, L. Moore, S. L. Adams,
A. Millar, P. Taylor, K. Bennett, K. Boutilier, L. Yang, C. Wolting,
I. Donaldson, S. Schandorff, J. Shewnarane, M. Vo, J. Taggart,
M. Goudreault, B. Muskat, C. Alfarano, D. Dewar, Z. Lin,
K. Michalickova, A. R. Willems, H. Sassi, P. A. Nielsen,
K. J. Rasmussen, J. R. Andersen, L. E. Johansen, L. H. Hansen,
H. Jespersen, A. Podtelejnikov, E. Nielsen, J. Crawford, V. Poulsen,
B. D. Sorensen, J. Matthiesen, R. C. Hendrickson, F. Gleeson,
T. Pawson, M. F. Moran, D. Durocher, M. Mann, C. W. Hogue,
D. Figeys and M. Tyers: Systematic identification of protein com-
plexes in Saccharomyces cerevisiae by mass spectrometry. *Nature*
415, 180 (2002).

[18] A. Hoffmann, T. Oelgeschlager and R. G. Roeder. Considerations of
transcriptional control mechanisms: Do TFIID-core promoter complexes recapitulate nucleosome-like functions? *Proc. Natl. Acad. Sci.
USA* **94**, 8928 (1997).

[19] M. J. Holland. Transcript abundance in yeast varies over six orders of magnitude. *J. Biol. Chem.* **277**, 14363 (2002).

[20] T. E. Ideker. Network genomics. *Ernst. Schering. Res. Found. Workshop.* 89 (2007).

[21] J. K. Joung, E. I. Ramm and C. O. Pabo. A bacterial two-hybrid selection system for studying protein-DNA and protein-protein interactions. *Proc. Natl. Acad. Sci. USA* **97**, 7382 (2000).

[22] M. Kanehisa and S. Goto. KEGG. Kyoto encyclopedia of genes and genomes. *Nucleic Acids Res.* **28**, 27 (2000).

[23] T. K. Kim and T. Maniatis. The mechanism of transcriptional synergy of an in vitro assembled interferon-beta enhanceosome. *Mol. Cell.* **1**, 119 (1997).

[24] R. Kodzius, M. Kojima, H. Nishiyori, M. Nakamura, S. Fukuda, M. Tagami, D. Sasaki, K. Imamura, C. Kai, M. Harbers, Y. Hayashizaki and P. Carninci. CAGE: Cap analysis of gene expression. *Nat. Methods* **3**, 211 (2006).

[25] M. Kozak. Initiation of translation in prokaryotes and eukaryotes. *Gene.* **234**, 187 (1999).

[26] M. Kunzler, C. Springer and G. H. Braus. The transcriptional apparatus required for mRNA encoding genes in the yeast Saccharomyces cerevisiae emerges from a jigsaw puzzle of transcription factors. *FEMS Microbiol. Rev.* **19**, 117 (1996).

[27] A. B. Lassar, P. L. Martin and R. G. Roeder. Transcription of class III genes: Formation of preinitiation complexes. *Science* **222**, 740 (1983).

[28] D. Lin, D. L. Tabb and J. R. Yates III. Large-scale protein identification using mass spectrometry. *Biochim. Biophys. Acta.* **1646**, 1 (2003).

[29] D. W. Lin and P. S. Nelson. Microarray analysis and tumor classification. *N. Engl. J. Med.* **355**, 960; author reply 960 (2006).

[30] R. X. Luo and D. C. Dean. Chromatin remodeling and transcriptional regulation, *J. Natl. Cancer. Inst.* **91**, 1288 (1999).

[31] N. M. Luscombe, M. M. Babu, H. Yu, M. Snyder, S. A. Teichmann, and M. Gerstein. Genomic analysis of regulatory network dynamics reveals large topological changes, *Nature* **431**, 308 (2004).

[32] M. Mann, R. C. Hendrickson and A. Pandey. Analysis of proteins and proteomes by mass spectrometry. *Annu. Rev. Biochem.* **70**, 437 (2001).

[33] S. Mathivanan, B. Periaswamy, T. K. Gandhi, K. Kandasamy, S. Suresh, R. Mohmood, Y. L. Ramachandra and A. Pandey. An evaluation of human protein-protein interaction data in the public domain. *BMC Bioinformatics* **7** (Suppl 5), S19 (2006).

[34] V. K. Mootha, C. M. Lindgren, K. F. Eriksson, A. Subramanian, S. Sihag, J. Lehar, P. Puigserver, E. Carlsson, M. Ridderstrale, E. Laurila, N. Houstis, M. J. Daly, N. Patterson, J. P. Mesirov, T. R. Golub, P. Tamayo, B. Spiegelman, E. S. Lander,

J. N. Hirschhorn, D. Altshuler and L. C. Groop. PGC-1alpha-responsive genes involved in oxidative phosphorylation are coordinately downregulated in human diabetes. *Nat. Genet.* **34**, 267 (2003).

[35] R. Nilsson, V. B. Bajic, H. Suzuki, D. di Bernardo, J. Bjorkegren, S. Katayama, J. F. Reid, M. J. Sweet, M. Gariboldi, P. Carninci, Y. Hayashizaki, D. A. Hume, J. Tegner and T. Ravasi. Transcriptional network dynamics in macrophage activation. *Genomics* **88**, 133 (2006).

[36] T. Oelgeschlager, C. M. Chiang and R. G. Roeder. Topology and reorganization of a human TFIID-promoter complex. *Nature* **382**, 735 (1996).

[37] S. Peri, J. D. Navarro, R. Amanchy, T. Z. Kristiansen, C. K. Jonnalagadda, V. Surendranath, V. Niranjan, B. Muthusamy, T. K. Gandhi, M. Gronborg, N. Ibarrola, N. Deshpande, K. Shanker, H. N. Shivashankar, B. P. Rashmi, M. A. Ramya, Z. Zhao, K. N. Chandrika, N. Padma, H. C. Harsha, A. J. Yatish, M. P. Kavitha, M. Menezes, D. R. Choudhury, S. Suresh, N. Ghosh, R. Saravana, S. Chandran, S. Krishna, M. Joy, S. K. Anand, V. Madavan, A. Joseph, G. W. Wong, W. P. Schiemann, S. N. Constantinescu, L. Huang, R. Khosravi-Far, H. Steen, M. Tewari, S. Ghaffari, G. C. Blobe, C. V. Dang, J. G. Garcia, J. Pevsner, O. N. Jensen, P. Roepstorff, K. S. Deshpande, A. M. Chinnaiyan, A. Hamosh, A. Chakravarti and A. Pandey. Development of human protein reference database as an initial platform for approaching systems biology in humans, *Genome. Res.* **13**, 2363 (2003).

[38] J. Quackenbush. Extracting biology from high-dimensional biological data. *J. Exp. Biol.* **210**, 1507 (2007).

[39] B. Ren, F. Robert, J. J. Wyrick, O. Aparicio, E. G. Jennings, I. Simon, J. Zeitlinger, J. Schreiber, N. Hannett, E. Kanin, T. L. Volkert, C. J. Wilson, S. P. Bell and R. A. Young. Genome-wide location and function of DNA binding proteins. *Science.* **290**, 2306 (2000).

[40] J. F. Rual, K. Venkatesan, T. Hao, T. Hirozane-Kishikawa, A. Dricot, N. Li, G. F. Berriz, F. D. Gibbons, M. Dreze, N. Ayivi-Guedehoussou, N. Klitgord, C. Simon, M. Boxem, S. Milstein, J. Rosenberg, D. S. Goldberg, L. V. Zhang, S. L. Wong, G. Franklin, S. Li, J. S. Albala, J. Lim, C. Fraughton, E. Llamosas, S. Cevik, C. Bex, P. Lamesch, R. S. Sikorski, J. Vandenhaute, H. Y. Zoghbi, A. Smolyar, S. Bosak, R. Sequerra, L. Doucette-Stamm, M. E. Cusick, D. E. Hill, F. P. Roth and M. Vidal. Towards a proteome-scale map of the human protein-protein interaction network, *Nature* **437**, 1173 (2005).

[41] M. Schena, D. Shalon, R. W. Davis and P. O. Brown. Quantitative monitoring of gene expression patterns with a complementary DNA microarray. *Science* **270**, 467 (1995).

[42] C. A. Spencer and M. Groudine. Transcription elongation and eukaryotic gene regulation. *Oncogene.* **5**, 777 (1990).

[43] A. Subramanian, P. Tamayo, V. K. Mootha, S. Mukherjee, B. L. Ebert, M. A. Gillette, A. Paulovich, S. L. Pomeroy, T. R. Golub, E. S. Lander and J. P. Mesirov. Gene set enrichment analysis: a knowledge-based approach for interpreting genome-wide expression profiles, *Proc. Natl. Acad. Sci. USA* **102**, 15545 (2005).

[44] H. Suzuki, R. Saito, M. Kanamori, C. Kai, C. Schonbach, T. Nagashima, J. Hosaka and Y. Hayashizaki. The mammalian protein-protein interaction database and its viewing system that is linked to the main FANTOM2 viewer. *Genome. Res.* **13**, 1534 (2003).

[45] T. Tamura, Y. Makino and T. Kishimoto. [Regulation of gene expression and recent advance on transcription studies]. *Nippon Rinsho.* **53**, 1033 (1995).

[46] K. Tan, J. Tegner, and T. Ravasi. Integrated approaches to uncovering transcription regulatory networks in mammalian cells. *Genomics.* **91**(3), 219–231, Epub 2008 Jan 8 Review., (2008).

[47] J. Tegner, R. Nilsson, V. B. Bajic, J. Bjorkegren and T. Ravasi: Systems biology of innate immunity, *Cell. Immunol.* **244**, 105 (2006).

[48] T. Tsukiyama and C. Wu. Chromatin remodeling and transcription. *Curr. Opin. Genet. Dev.* **7**, 182 (1997).

[49] S. C. Tucker and R. Wisdom. Site-specific heterodimerization by paired class homeodomain proteins mediates selective transcriptional responses. *J. Biol. Chem.* **274**, 32325 (1999).

[50] M. W. Van Dyke, M. Sawadogo and R. G. Roeder. Stability of transcription complexes on class II genes. *Mol. Cell. Biol.* **9**, 342 (1989).

[51] C. L. Wei, Q. Wu, V. B. Vega, K. P. Chiu, P. Ng, T. Zhang, A. Shahab, H. C. Yong, Y. Fu, Z. Weng, J. Liu, X. D. Zhao, J. L. Chew, Y. L.Lee, V. A. Kuznetsov, W. K. Sung, L. D. Miller, B. Lim, L. Lee, E. T. Liu, Q. Yu, H. H. Ng and Y. Ruan. A global map of p53 transcription-factor binding sites in the human genome. *Cell.* **124**, 207 (2006).

Chapter Thirteen

Gene-Expression Ontologies and Tag-Based Expression Profiling

Oliver Hofmann[1,*] and Winston Hide[1,2]

[1] *Department of Biostatistics, Harvard School of Public Health, USA*
[2] *South African National Bioinformatics Institute (SANBI), University of the Western Cape, South Africa*
*Email: *ohofmann@hsph.harvard.edu*

With the advent of the next generation sequencing and high-throughput technologies such as CAGE, a large amount of data is being generated. However, in absence of well organized databases and ontologies, it becomes difficult to retrieve and understand biological significance across experiments and datasets. Here we discuss these issues, particularly in relation to tag-based expression profiling approaches like CAGE.

13.1 INTRODUCTION

Discoveries that have resulted from the study of RNA that has been expressed from the genome — transcriptomics — have been dependent upon the availability of technologies that can consistently capture a representation of expressed RNA. Sampling technologies for transcripts have diversified after two gateway technologies: RNA northern analysis and Reverse Transcriptase (RT) PCR and Quantitiative RT–PCR studies. These in turn have become verifying approaches for the now ubiquitous RNA microarray platforms. High sensitivity detection of gene expression products can now be performed with increasingly lower amounts of starting RNA. The principal drawback of these

Cap Analysis Gene Expression (CAGE): The Science of Decoding Gene Transcription **edited by P Carninci**
Copyright © 2010 by Pan Stanford Publishing Pte Ltd
www.panstanford.com
978-981-4241-34-2

ubiquitous probe or primer based sampling technologies is that they sample what is known already. Only transcripts matching to predetermined array probes will be detected. In addition, the relationship between a hybridization signal from an array is an unknown in terms of the labelled RNA: a relationship must be assumed by means of a canonical interpretation of the probes being used.

Transcript sampling approaches independent of a set of predefined assay probes, such as Massively Parallel Signal Sequencing (MPSS), Serial Analysis of Gene Expression (SAGE) or Expressed Sequence Tag (EST) sequencing, are powerful in that they have breadth of transcript coverage and provide insight into novel transcript expression events. However these technologies typically do not sample the transcriptome very deeply. Recently, "next generation" sequencing technologies have begun to be applied to whole transcriptomes, providing in depth short sequence read representations of transcriptional events.

CAGE approaches have also benefited by the availability of low-cost deep sampling technologies using the SOLEXA, SOLiD and 454 platforms. The key attributes of CAGE read data remain unchanged: (a) CAGE reads present a "snapshot" of transcription initiation at a genomewide scale. (b) The initiation snapshot has internal consistency in terms of stoichiometry: manipulation of reads matching known promotors can yield relative expression estimates for known genes. (c) The understanding of the exisitence of non-coding gene expression has been opened up dramatically by the availability of CAGE tags to genome matches that have provided new insight into non-coding RNA initiation.

As in all RNA data, a record of the source of the RNA libraries sampled provides insight into expression events captured by the library. Using a comprehensive approach to cataloging and structuring RNA library descriptions, we have developed a resource for the description of gene expression across expression states and technology platforms.[1] This resource now incorporates the current public CAGE corpus. As of February 2008 the CAGE library collection (Genome Network Version, http://genomenetwork.nig.ac.jp/public/download/cage_Data base_e.html) contains samples from 103 different human and 209 mouse tissue samples, covering a wide variety of individual tissues, pathological and developmental states (see Table 1) as well as a number of chemical treatments. Biological samples include a panel of normal tissues (such as different brain regions, thymus, muscle, testis, among others) from both adult and embryonic or fetal samples, paired tumour/stroma samples from a number of different patients (rectal cancer, kidney cancer, intestinal cancer, hepatoma, pancreatic tumours, melanoma) as well as a collection of additional CAGE libraries generated from cell lines (lymphoma, somatic stem cells) allowing for a comparison across different biological states of highly expressed genes after conversion of absolute tag counts to normalized tag-per-million counts.

With a total of twelve million sequenced human CAGE tags in the public database, current estimates of transcription start site numbers range from a conservative 30.000 to more than 100.000 sites depending on the minimum number of overlapping tags required before making a transcription start site (TSS) call (at least 5 tags or at least 20 tags at one start site respectively). See also Chapters 14 and 15 for a broader discussion of core promoter identification. As new CAGE libraries get added to the collection we

Table 13.1 (a) Five CAGE libraries each with the highest tag count from normal, developmental or tumor tissues. (b) Overview of CAGE library and cumulative tag counts from normal, developmental, tumor and stem cell tissues. The 'other' group includes cell lines as well as chemically modified tissue culture samples.

Group	Tissue	Tags
Normal	occipital lobe	576.277
	liver	244.260
	parietal lobe	226.352
	testis	154.865
	small intestine	150.277
Developmental	brain	339.886
	whole body	287.503
	lung	150.144
	spleen	128.396
	colon	106.907
Cancer	liver	2000022
	large intestine	498.083
	cecum	449.153
	bone marrow	387.724
	colon	269.288

(a)

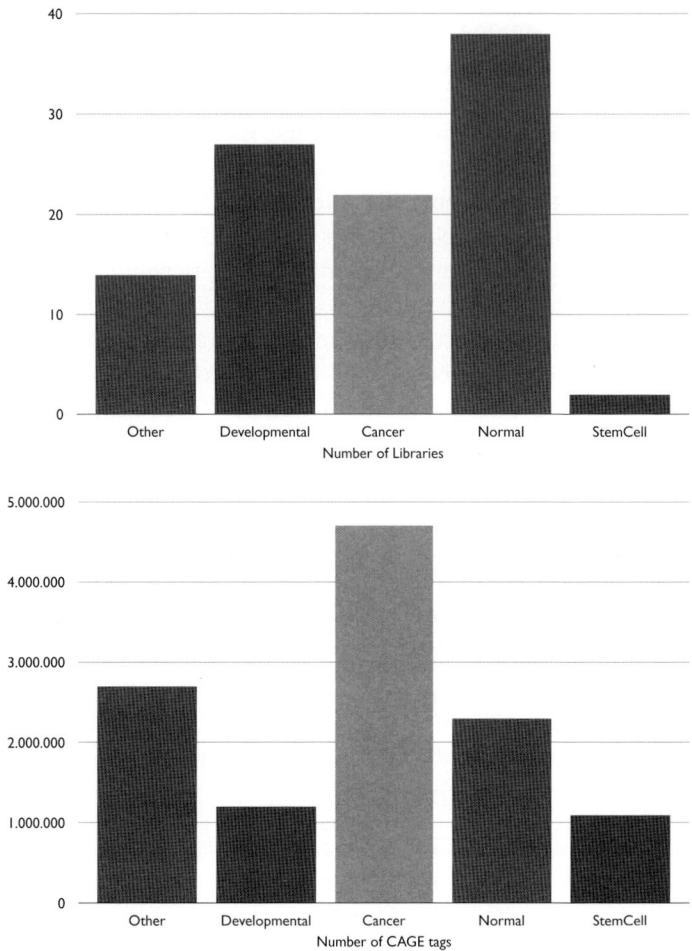

(b)

continue to see new and distinct start sites unique to particular gene expression states, indicating we are not yet at saturation with regards to the identification of novel start sites and reiterating the need for additional samples from as many tissues, pathological and developmental states as possible to complete our inventory of start site utilization. Additional libraries will also improve our ability to group high-confidence TSS into more well-defined subgroups beyond the current broad classes of 'normal' or 'cancer' transcript initiation.

13.2 ANNOTATING GENE EXPRESSION

Ontologies — structured vocabularies organized in formalized relationships such as 'part-of' or 'develops-from' — are commonly used to describe biological features.[2] Simply by applying a commonly accepted term such as 'heart' with a precise, well-defined and community accepted definition to the biological source descriptions of expression libraries, those libraries gain a common descriptor which can then be utilized to identify all instances of heart-associated expression information (e.g., those created from fetal heart or an adult ventral chamber). Thus, when seeking to relate "heart" to other terms, or to find out what is expressed in the parent term "heart" an investigator can manipulate understanding of gene expression in silico — in a controlled manner. This approach is dependent upon verified and curated relationships between the terms themselves (parent / child relationships, see Fig. 13.1) and biological objects such as libraries; a curator must have reviewed and assigned a term-library relationship. In this manner, we have applied the eVOC ontology system to the CAGE libraries available through the Genome Network Platform.

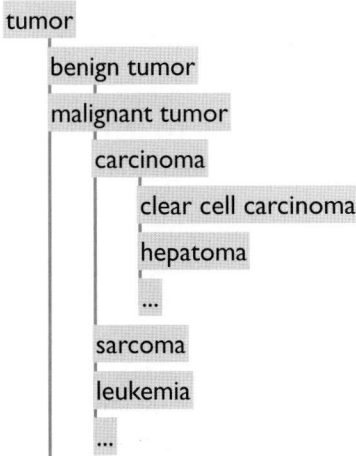

Figure 13.1. Sample excerpt of the eVOC tumor hierarchy, part of the controlled vocabulary to describe the pathological status of a source tissue. See http://evocontology.org for the full subset.

The digital readout of expression strength generated by CAGE data provides an additional benefit for integration purposes: TSS that might be too weak to reliably call based on CAGE tags from a single library can be identified and characterized due to individual contributions from all libraries within the collection. While a single tag — even when mapped uniquely to a distinct genomic location — is not sufficient to identify a transcription initiation a group of tags from different biological samples all mapping to the same area, overlapping each other and ideally with a distinct peak highlighting the most likely initiation site is in most cases sufficient, circumventing otherwise existing problems with low tag coverage in selected CAGE libraries. Effectively, this also increases the TSS count for most gene expression states as TSS no longer need to be defined by a single CAGE library but can be "annotated" as being active in a given tissue or disease state if more than N tags contribute to a globally identified start site. While alternate promoters unique to a particular library will most likely be missed, the approach can be helpful if, for technical reasons, only a shallow sequencing is possible or the initiation site of a gene weakly expressed in one tissue is identical with the start site in tissues with higher expression. Start sites determined from a few selected libraries that have been sequenced for high coverage can then be used to impute start sites most likely also active in a larger number of samples sequenced at much lower resolution, trading depth of a study for larger sample variation.

13.3 USING ONTOLOGIES TO INTEGRATE EXPRESSION INFORMATION

Given that all CAGE libraries are neither normalized or subtracted, an integration across different libraries to obtain a higher-level representation of transcription start site utilization and expression strength is straightforward provided all libraries are fully annotated with a hierarchical vocabulary or ontology. Combining all libraries within an ontology subtree for comparison with other subtrees enables a broader coverage of genes and non-coding RNAs beyond what would be possible by a pairwise comparison of individual libraries (see Fig. 13.2). Instead of contrasting CAGE tags from TSS associated with a liver tumour and matching stroma samples it is possible to observe more general

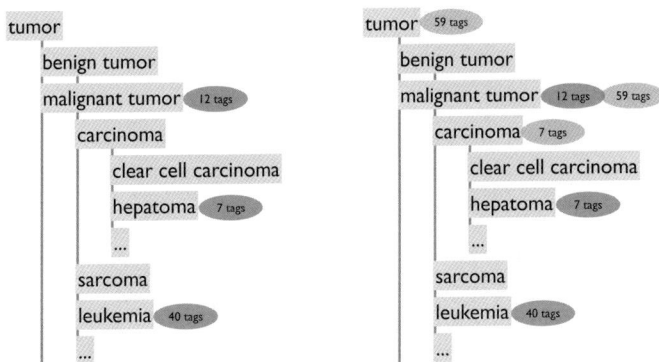

Figure 13.2. Propagation of CAGE tag counts based on specific library annotations to more generic term, allowing for the grouping of CAGE tag information — and transcription start sites — into broader biological classes. Original counts on the left hand side in green, cumulative tag counts in orange on the right hand side.

changes such as similarities and differences between all TSS active in neoplastic tissues and those found to be active in normal ones.

More generally, any level of abstraction can be applied to the underlying CAGE tag distribution by not only looking at differences between samples from normal and pathological tissues, but to also look for differences in tissue distributions — what start sites are highly selective for a specific tissue — or differences in distribution during developmental stages (see Fig. 13.3). Which TSS are specific to developmental tissues and what start sites seem to be shared across all stages of fetal, embryonic and adult tissues? Start site information grouped by their activity in different gene expression states can then be analyzed for shared regulatory motifs or other organizing principles.

As the defining characteristic of CAGE tag is their mapping to one or more genomic locations it is possible to combine tag and TSS distributions with other transcriptome resources such as the 5'- end-sequences from dbTSS 3 (Reference dbTSS: DBTSS, DataBase of Transcriptional Start Sites: progress report 2004. by: Y Suzuki, R Yamashita, S Sugano, K Nakai. Nucleic Acids Res, Vol. 32, No. Database issue. (1 January 2004) or full length cDNA libraries that identify transcription initiation sites on an identical genome build. At the most basic level a simple merger of overlapping TSS provides additional confidence in the quality

TSS classification

chr1: | 154910000   154915000

Normal
Developmental
Cancer

NES

chr1: 154909100 154909200 154909300 154909400 154909500 154909600 154909700 154909800
Cancer:HBY

Developmental:HDS   Cancer:HCB
Developmental:HDS   Cancer:HCB
Developmental:HDS   Cancer:HCB
Developmental:HDS   Cancer:HCB
Developmental:HDS   Cancer:HCB
Developmental:HDS   Cancer:HCB
Developmental:HDS
Developmental:HDS
Developmental:HDS
Developmental:HDS
Developmental:HDS
Developmental:HDS
Developmental:HDS
Developmental:HDS
Developmental:HDS
Developmental:HDS
Developmental:HDS
Developmental:HDS
Developmental:HDS

Developmental
Cancer

NES

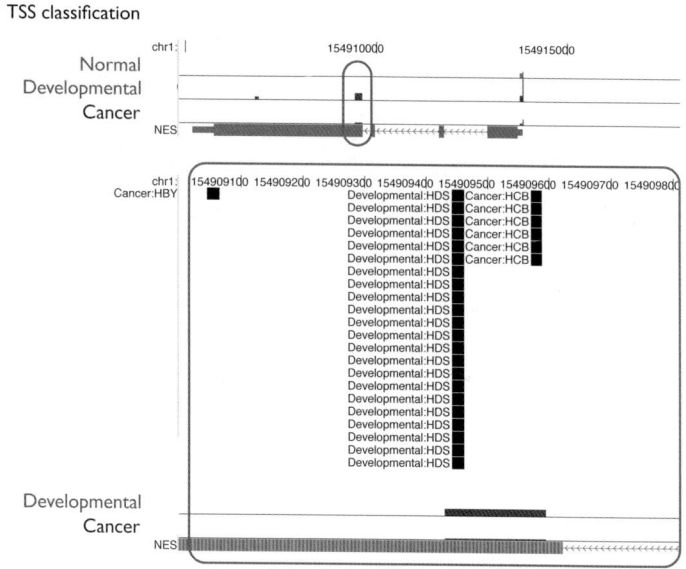

Figure 13.3. (a) Sample CAGE tag distribution for Nestin (NES, Entrez Gene Identifier 10763) across all libraries generated from normal (green), developmental (cyan) or tumor (red) samples resulting in calls for transcription start sites. (b) Close-up of the 154,909,000 region of human chromosome 1 (highlighted in red), showing the position of CAGE tags from a developmental lung library (HDS) and one neuroblastoma cancer library (HCB) in close proximity. A single tag from a hepatoma library (HBY) was insufficient to result in the call of a transcription start site.

of the mapping and allows to disambiguate multiple mappings of CAGE tags based on evidence obtained from other expression platforms. More sophisticated approaches only link promoter structures in close proximity if evidence of their activity is based on expression libraries with similar biological properties (i.e., sharing the same ontological annotation) as to avoid merging potentially unique initiation sites into an overly broad representation. In addition, the digital readout of tag-based technologies can be translated into a tags-per-million representation, allowing for a direct integration of normalized expression information across platforms, which is difficult to achieve with intensity-based array platforms. As an example, we have used the eVOC system to organize CAGE, EST and cDNA resources so that they can

Testis   Normal, adult tissues

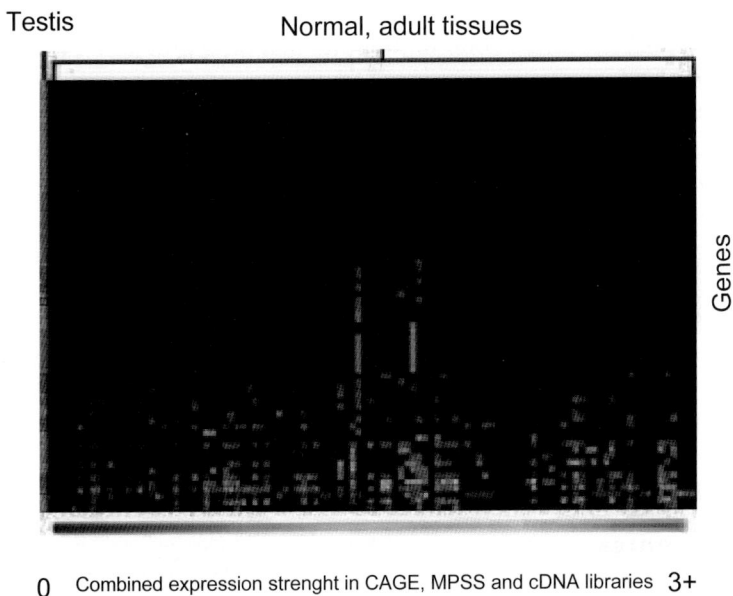

0   Combined expression strenght in CAGE, MPSS and cDNA libraries   3+

Figure 13.4. Set of X-chromosomal genes (rows) exhibiting testis-specific expression in a collection of CAGE-, MPSS- and cDNA-libraries. Each column represents a normal adult tissue for which expression libraries were available. Color reflects the normalized expression strength of the gene in a given tissue, with each expression source contributing equal parts to the final expression score, and genes are ranked by the ratio of expression strength in testis compared to all other normal tissues.

readily be interrogated by developmental stage, pathology, and anatomy (as well as species) to screen for genes on the human X-chromosome exhibiting a bias for a testis-restricted expression (see Fig. 13.4 and Ref. 3).

The integration of expression information across heterogeneous data sources comes with a number of caveats. One of the key issues in integrating CAGE-based expression information with data from other platforms revolves around the concept of transcripts. Mapping CAGE tag clusters to downstream genes gets less reliable with distance, and unless matching array data is available to determine gene expression patterns matching a given TSS over time we are usually limited to matching a conservative, small window immediately upstream of a known, annotated gene region to this gene. In addition, identifying which alternate transcript might be associated with a given alternate promoter is not possible, further confounding the mapping problem. Finally, while EST sequences are mapped to a single gene (the 'best' match, e.g. using UniGene clusters) the 20–27 bp CAGE tags are frequently multi-mapped to similar genomic regions. While this increases sensitivity and allows us to identify regions of interest it comes at a cost of specificity, even considering a rescue approach that increases the weight of tags associating to regions that also contain uniquely mapped sequence reads (Chapter 8). Combining these considerations with frequency count issues prevalent in EST libraries which often are normalized or subtracted it becomes difficult to quantitatively compare expression information across multiple platforms. A qualitative test for presence or absence of expression, however, is very much feasible.

References

[1] Ashburner *et al.* Gene ontology: Tool for the unification of biology. The Gene Ontology Consortium. *Nat. Genet.* **25**(1), 25–29 (2000).

[2] Kelso *et al.* eVOC: A controlled vocabulary for unifying gene expression data. *Genome Res.* **13**(6A), 1222–1230 (2003).

[3] Kruger *et al.* Simplified ontologies allowing comparison of developmental mammalian gene expression. *Genome Biol.* **8**(10) R229 (2007).

Chapter Fourteen

Lessons Learned from Genomic CAGE

A. Sandelin* and E. Valen[†]

Department of Biology & Biotech Research and Innovation Centre, University of Copenhagen, Denmark
*Email: *albin@binf.ku.dk, [†]eivind@binf.ku.dk*

In this chapter, we describe some of the major biological lessons learned up to now by applying CAGE to the human and mouse genomes. We will contrast these finding to text-book views on core promoters and transcription initiation, and in the end show that the data opens many doors for both experimental and computational analysis of core promoters and their mechanisms.

14.1 INTRODUCTION

In pre-genome bioinformatics — around 2000 AD — there was a belief that the completion of the human genome project would boost the demand for sophisticated prediction methods for genes, promoters and protein structures, since at the time all techniques used to investigate these features were laborious even for single loci. While this was partially correct, the major sources of data to understand genomes now originate from actual experiments, by the use of novel high throughput techniques either based on sequencing of full or partial transcripts, or hybridizing transcripts to tiling arrays. Since all of these techniques are at least partially dependent on accurate genome assemblies, the speed in which these techniques have gone from concepts to maturation

Cap Analysis Gene Expression (CAGE): The Science of Decoding Gene Transcription **edited by P Carninci**
Copyright © 2010 by Pan Stanford Publishing Pte Ltd
www.panstanford.com
978-981-4241-34-2

is noteworthy. This means that the field of transcriptomics, and genomics in general, now is driven by the experimental methods and their applications on whole genomes, which in turn demands computational techniques for analysis.

One of the returning lessons of large scale transcriptome annotation projects using new experimental technologies is the amount of unexpected genome features that are discovered due to that the methods used are unbiased in terms of target transcripts, compared to targeted "classical" molecular biology. This indicates that there is a substantial ascertainment bias in targeted experiments molecular biology.

The findings made by applying the CAGE technology to genomes is one of the strongest examples of this, because these data is changing the text-book views of transcription start sites that most biologists carry with them from their undergraduate studies. In this chapter, we have tried to broadly summarize the lessons learned from CAGE so far, and also highlight some examples of studies that were motivated from these findings. We conclude by pointing out some directions and challenges for future work on understanding transcription initiation.

14.2 THE CLASSIC VIEW ON TRANSCRIPTION START SITES AND CORE PROMOTERS

Much of what we know of the mechanisms and properties of core promoters dates back to the pioneers of promoter analysis. For natural reasons, the types of promoters that were the focus of investigation were strongly used by cells and easily inducible — many of them were tissue-specific. Almost all the eukaryotic promoters studied had a TA-rich region, (named the TATA-box after its consensus sequence) starting some 30 bp before the point of transcription initiation. It was then natural to assume that most, if not all, promoters were governed by such a TATA box, and that the core promoter could be regarded as a template which always had the same functional content — thus, the most interesting and variable regulatory elements resided elsewhere.[1] This fitted very well with the fact that the TATA-binding protein is a part of the initiation complex, and absolutely required for transcription using RNA Polymerase II.

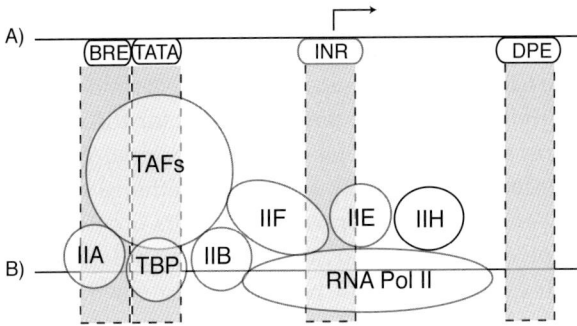

Figure 14.1. Simplified canonical view of core promoters and the assembly of the initiation complex, assuming a single TSS. Although DNA here is shown as a straight line, the TATA-box binding protein TBP bends DNA in a sharp angle when bound to the TATA-box. (a) Some of the known core patterns and their position relative to the TSS; note that several other patterns are known. (b) Parts of the initiation complex, which primes RNA Polymerase II to start transcription. The assembly begins with TFIID binding to the TATA box by its largest subunit — TBP.

As more promoters were studied, the field had to revise this view — there were clearly many promoters lacking the TATA-box, and several other sequence patterns reoccurring in core promoters with proven effect were discovered, including the Inr, BRE and DPE elements among others (see Smale et al.[1] for an excellent review). Figure 14.1 shows the most studied promoter elements as well as the major part of the initiation complex. At the same time, several core promoters with multiple points of initiation were discovered.

Regardless of this, the TATA-box-dominated promoters with a single point of initiation became the typical textbook picture describing promoters, likely because it is a simple way of explaining the promoter mechanism. Analogously, the only reasonably detailed models we have for transcription initiation assumes that the TATA-binding protein is recruited by a viable TATA-box, and that initiation occur at one place only (reviewed in Ref. 2).

The focus of the above model is to understand mechanistically how initiation takes place. In computational biology, a variant of this problem is how to *predict* promoters from genome sequence, with the goal of annotating genomes without time-consuming experiments. While predictive tools do not necessarily have to be based on biological knowledge, most of the significant algorithms developed employed a mix of biologically known patterns as well as machine learning techniques, and were also trained on promoters fitting to the single TSS concept above, for the simple reason that the most studied promoters were of this type. With these definitions, promoter finding as a field has hitherto not been very successful — in fact, many promoter finders are essentially CG-dinucleotide detectors,[3] motivated by the high degree of overlap of between CpG-islands and promoters. In all, the accuracy of such programs were not sufficient to annotate genomes, and this was one of the major motivations to develop high-throughput TSS detection assay such as CAGE (the other motivation was to assess actual expression levels of individual promoters). In the end of Section 4 we will return to the promoter prediction problem and some new avenues for solving it, or at least re-define it, made possible by CAGE data.

14.3 CAGE-BASED VIEWS OF TRANSCRIPTION START SITES

In this section, we will recapitulate a number of basic findings of the nature of core promoters enabled by CAGE, and contrast these to the classic view discussed above. We are deliberately avoiding discussing subjects that have other devoted chapters in this book in any depth, for instance promoter evolution. In Section 4, we will extend on these findings by reviewing some more detailed studies based on this.

14.3.1 *Transcription Start Site Landscapes*

One of the largest surprises when mapping CAGE tags to the mouse and human genomes was that in most cases, the mapped tags did not conform to the classical picture of a sole TSS.[4,5] In fact, the tags indicated a dense forest of TSSs, where some nucleotides were used more than others. Figure 14.2 shows examples of this from both human and mouse promoters.

   This phenomenon was first assumed to be due to experimental noise; therefore, tags that were overlapping each other on the same strand were merged into a single entity: a tag cluster. On closer inspection of these tag clusters, it quickly became apparent

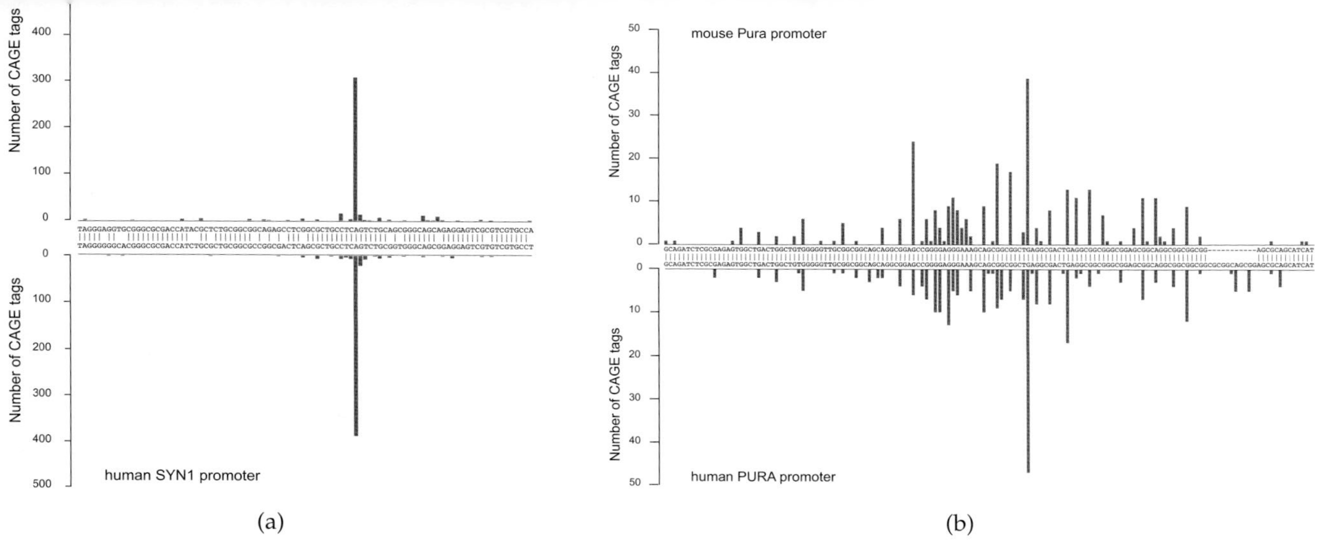

Figure 14.2. Simplified canonical view of CAGE tag count distributions in human and mouse. In these images, the number of times a particular nucleotide is used as a 5′ end is shown at the y-axis. The x-axis is an alignment of human and mouse genome sequences, where the upper part of each pael is mouse. Transcription is right to left. (a) The core promoter for the *Syn1* gene shows many of the characteristics of a text-book promoter: almost all initiation events occur at a single nucleotide. Note that this is captured in both human and mouse. Worth to note is that this promoter does not have a clear TATA box, only an Inr box just at the main initiation peak. (b) The *Pura* promoter displays a different distribution of tags — many initiation events over a wide genomic space, mirrored in human and mouse. This type of distribution is more representative than the single peak type in (a) for promoters with enough tags, even though it should be noted that many different types of distributions can be observed.

that noise was not a good explanatory model to the TSS distributions within the clusters, due to three reasons:

(1) The classical type of promoters — a single TSS often governed by a TATA-box, could be found reasonably often using the CAGE tags — in such cases, almost all tags were mapping to one position, regardless of tissue origin. This is in itself an argument for the validity of CAGE, because if more shallow CAGE tag distributions would be due to experimental noise, then these single position distributions would be extremely unlikely to occur. The *Syn1* core promoter in Figs.14.2(a) and 14.3(b) is a good example of this.

(2) Different types of CAGE tags distributions occur within one and the same CAGE library and, as pointed out above, the distributions are consistent for a given core promoter (although shifts can occur at times, see below). If random noise had a great impact, we would expect that all core promoters had a Gaussian CAGE tag distribution, or at least that the different libraries had different types of distributions in each promoter (see Fig. 14.3)

(3) When comparing orthologous mouse and human promoters, the CAGE tag distributions were remarkable similar, despite that few of the CAGE tag libraries in the two species could easily be paired (see Fig. 14.2)

When comparing well-annotated promoters with multiple TSSs, these generally agree well with the CAGE data — differences are often due to that the CAGE tag sampling is much deeper than that of the original experiments. In fact, when re-examining these older studies we experienced that the nuclease protection assays used often indicated arrays of weaker TSSs as smears in the gel pictures, which the authors at the time interpreted as noise. This was also discussed in the introduction section — many well-characterized promoters have several TSSs, and other semi-high-throughput studies also corroborate these result.[6,7] The biological merit of the different distributions motivated a hierarchal promoter classification system based on the TSS distribution. At the highest level it was natural to divide between the classical promoters, with strong spatial restrictions, and the broader majority type. Carnici *et al.*[5] proposed a system where a tag cluster is defined as "single peak" if it contains 75% of its tags within a 4bp region and "broad" otherwise, and then used sub-classes of

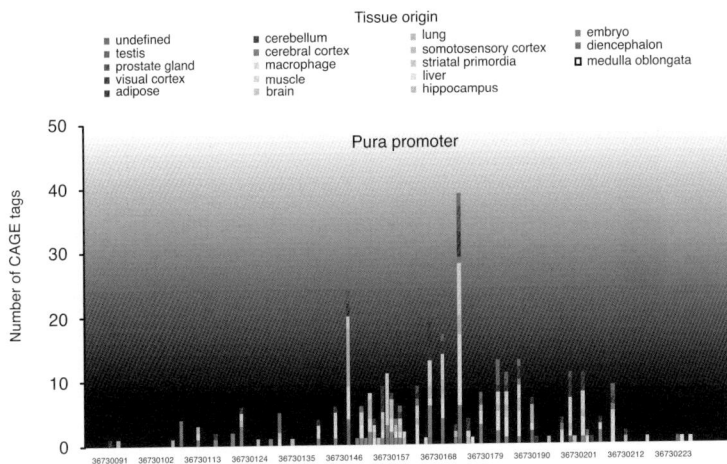

(a)

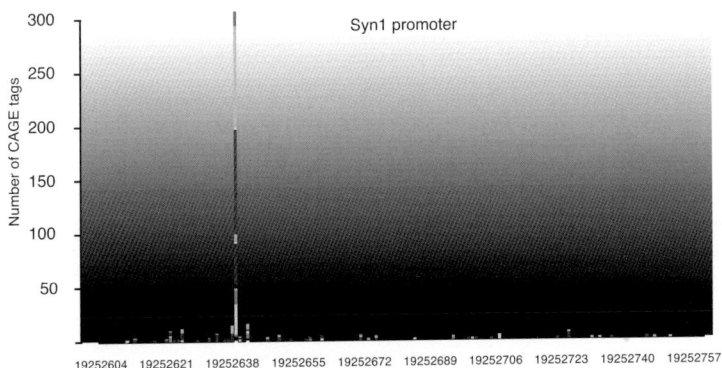

(b)

Figure 14.3. Tissue usage of core promoters. (a) Detailed view of the *Pura* mouse promoter from Fig. 14.2(b) where the tag contribution from different tissue are shown. Note that all libraries are not of equal total size, but the general trends in the distribution trend is present in mouse tissues. (b) Detailed view of the *Syn1* mouse promoter from Fig. 14.2(a) (note that the direction is reversed due to genome coordinate labeling consistency). Again, the sharp peak is composed of tags from multiple tissues. (c) Example tissue usage shift in a core promoter. Only liver and lung tags in the mouse *Ppap2b* promoter are shown: while the first peak is used by both tissues, the second is dominated by lung.

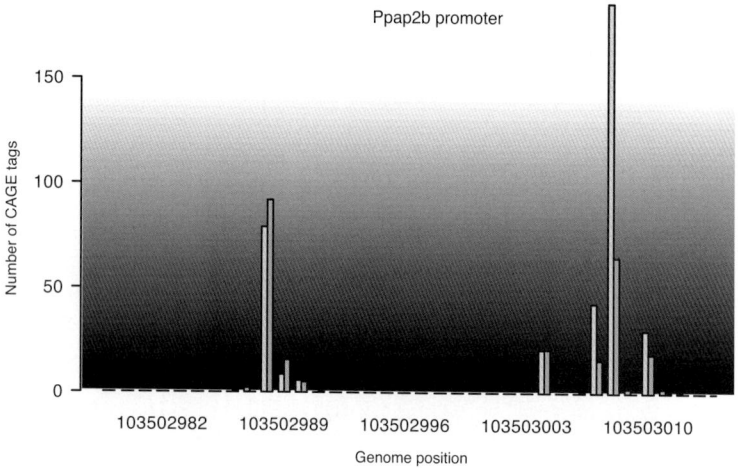

Figure 14.3(c)

the broad category. While this classification is somewhat arbitrary, it has proved useful in exploring the many differences in promoter architecture. Many of these are discussed below.

14.3.2 *Biological Functions of TSS Distributions*

In general, the shallow, broad CAGE distributions were over-represented in CG-rich patterns (CpG islands in particular), whereas the distribution with a narrow single peak were, as expected, over-represented in TATA-boxes.[5] While the over-representations were highly significant, it is worth to point out that all individual exceptions are common — there are single peak promoters with CpG islands and no TATA-boxes and broad promoters without CpG islands, for instance. Furthermore, single peak promoters were biased towards a certain tissue more often than the shallow peak. This is consistent with the link between CpG islands and ubiquitous genes on one hand, and TATA-boxes and tissue specific genes on the other. As above, there are some clear exceptions to this: for instance, promoters preferentially used in brain or in embryonal development are often CG-rich.

Using CAGE data, Taylor *et al.*[8] showed that CpG island promoters evolve significantly faster than TATA-box promoters (see also Chapter 16). This would indicate that the shallow TSS distributions have a higher evolutionary plasticity, which makes sense since single point mutations in these promoters would at most lead to an increase or decrease of the usage of one or a few TSSs. Conversely, point mutations in a single peak promoters would likely have a binary effect — either no change or a drastic decrease in expression, which likely would be deleterious. In other words, we postulate that one of the reasons for having multiple TSSs in the same promoter is to have an expression template that can be subtly modified during evolution.

14.3.3 *Alternative TSSs*

Another surprise from the high-throughput TSS sequencing was the sheer number of strong promoters inside known genes. From the initial study,[5] over 50% of known genes had two or more core promoters that were distinct, which is consistent with other studies.[9,10] This number is likely much larger, since the sampling was limited (not all genes are hit by CAGE tags, and all genes are not identified). Interestingly, if a given gene has more than one core promoter, these often display significantly different tissue expression preferences, which makes it likely that they are differentially regulated. So, the function of these alternative promoters can both be to be able to regulate the gene differently depending on context, but also to complement alternative splicing in generating different mRNA products (and even different proteins). While this sounds straightforward, it is often hard to judge the exact boundaries between core promoters, especially since many singleton tags are seemingly randomly distributed within exons of some genes (see below). This motivates more complex algorithms for clustering tags.

14.3.4 *Promoters at Unexpected Locations*

With a sampling as deep as the initial CAGE study (about 12 million tags in a mouse), it was expected to find many novel core promoters, especially for novel non-coding RNA in intergenic regions. While this was true, we will focus on two observations of core promoters within genes. The reason for this is that novel promoters with no other overlapping data are hard to characterize in terms of function without further experiments.

14.3.4.1 Weak Exonic Promoters

Above, we have discussed clusters of CAGE tags, but in fact, with the current sequencing depth, most CAGE tags are solitary. These singletons were removed from most analyses since they represent weak promoters and have a higher chance of being noise. From visual inspection, we saw that many of these singletons were distributed inside exons of known genes. Initially, this was though to be due inability of the reverse transcriptase to reach the 5' end, in combination with faulty capping. However, if this was the sole explanation, we would expect an exponential decrease of singleton occurrence from the 3' end of the gene (when using polyDT primers), which was not observed. At the same time we observed that some genes with strong 5' end promoters were devoid of these weak exonic TSSs, while others were almost covered with them. Generally, tissue-specific genes were over-represented in the set of genes that had many weak exonic TSSs. Moreover, these phenomena were consistent in mouse and human genes.

We still have no clear leads on what the implications of these observations are: certainly, some of the tags might be experimental noise, but due to the above, it is hard to throw away all such cases. We will highlight a few complementary studies which may explain part of these observations:

TSSs for gene-bridging long transcripts. In the pilot Encyclopedia for DNA Elements (ENCODE) study,[13] one of the most surprising findings was that there are long transcripts that can bridge genes or even span several genes, often starting in the middle of a gene structure. The function of these are unknown, although one hypothesis is that it is a way of transferring information of chromatin structure at the start site to the downstream gene. Some of the weak TSSs observed in the CAGE data could be attributed to such roles, although one important caveat is that we do not know the length of these transcripts — this is an in-built limitation of the technique.

Initiation but not transcription? Chromatin immuno-precipitation data also give some interesting interpretations. In contrast to the ENCODE findings, Guenter *et al.*[14] showed that some initiation events do not give a full-length product, even though the "correct" epigenetic marks and RNA Polymerase II is bound. The authors focused on annotated (strong) core promoters, but this could of course also apply for weaker promoters, which means that many of them will not give a viable product. If this is widespread, it might be a problem if wanting to correlate promoter usage with mRNA levels; it is at present hard to know if this is cause for concerns as CAGE is measuring promoter usage by sequencing (full-length) mRNAs.

Recapping. Recently it was suggested[11] that cap structures might also arise as a results of post-transcriptional modifications. Mature mRNA is in this model subjected to cleaving and subsequent re-capping or a similar process that renders them sensitive to cap extraction methods. The end result is what appears to be capped segments that does not correspond to transcription initiation sites. One of the main arguments in favor of this model was that several CAGE tags were found to extend across introns. Assuming these were a product of real initiation events the resulting first exons would be improbably short and have very low splicing efficiency.[12] If this model is correct it is a product of an, as of now, unknown pathway.

14.3.4.2 Core Promoters in 3' UTRs

A byproduct of the investigation of exonic TSSs was the finding that some protein-coding genes have a higher aggregation of TSSs in the last 20% portion of its last exon — usually in the 3' UTR. This is connected to recent findings using tiling arrays, where transcripts overlapping termination sites were observed.[15] The tissue usage of these promoters did not consistently correlate with that of the corresponding 5' promoters, suggesting that they are independently regulated. In a limited reporter gene assay, the 100 bp upstream of such promoters were found to be enough to drive expression. At the same time, we found a very clear GGG signal at the TSS, where the last G is at position +1 (the TSS). This strongly suggests that the function of these transcripts is not directly coupled to the "home" gene. However, an interesting observation was that if the closest gene downstream of the "home" gene was located on the opposite strand, the distance between the genes was significantly smaller than if the two genes were on the same strand, suggesting a role where the 3' UTR promoter gives transcripts bridging the genes to form an antisense interaction with the downstream gene. However, at this stage we do not know the typical length of these transcripts — the current guess is that they are short, as they are picked up when using oligoDT primers and not random primers.

14.4 PROBING BIOLOGICAL MECHANISMS USING CAGE

In this section we will focus on highlighting some studies made possible with the CAGE data. The reason for this is to emphasize the "pushing the boundary"-effect made possible with new data, and also to inspire further studies of these types. We will go from detailed questions towards questions pertaining the genome-wide rules for transcription: the last section will point towards many new types of studies, as it changes the views on core promoters and promoter prediction. Note that we have avoided discussing the detailed studies made on evolution of promoters and macrophage biology, and more, which we view as equally important; however, these are covered in Chapters 16 and 17.

14.4.1 *Measuring the Effect of TATA-TSS Spacing*

A point we have been trying to make throughout this chapter is that most promoters do not have a single peak of initiation, and are not governed by TATA-boxes. However, this does not in any way invalidate the studies made by the early gene regulation biology pioneers; these promoters do exists and they are an important biological area of study. Ponjavic *et al.*[16] returned to this promoter type to see the CAGE data depth could give new perspectives. This was the first study where a large number, 784, of TATA-box promoters could be studied in parallel within the same experimental dataset. Ponjavic *et al.* could pinpoint the distance constraints between the TATA-box and the TSS, and show that the distance had a significant effect on both tissue specificity and the sequence patterns in the initiation site. This article shows that with the added data depth the CAGE data gives, it can be worthwhile to revisit known concepts and see to what degree they hold in a genome-wide setting. A further argument for additional studies of this type is that a large amount of the promoter features we know are based on detailed studies of just a few genes.

14.4.2 Dynamic TSS Selection

Kawaji *et al.*[17] noted that in many cases, promoters with shallow distributions of TSSs could be decomposed to more than one TSS distribution depending on tissues, where the center was shifted. Figure 14.3(c) shows the promoter of the mouse *Ppap2b* gene, and CAGE tags from liver and lung, whose distributions clearly have different center points. Around 30% of 8157 promoters displayed at least one such shift between two tissues, and around 43% of the promoters displayed some internal shifts (so that a sub-region is enriched or depleted for a given tissue); in all, around 50% have one or both of these types of shifts when all sampled tissues are considered. This is slightly perplexing since the underlying DNA sequence is the same. It can be argued that the shift could depend on interactions with different types of intermediary complexes, which are composed differently in the two cell types. Alternatively, the shift could be due to different chromatin states — either methylation of DNA or simply shifts of the nucleosome binding patterns. In fact, promoters with shifts are biased towards paternal and maternal imprinting. A valuable lesson from this study is the fluidity of promoters — it strongly argues against seeing promoters as static blocks that are either used or not used, and is thus in sharp in contrast to most promoter prediction models. This also exemplifies how hard it is to cluster CAGE tags to functional units.

14.4.3 Evolutionary Turnover of Core Promoters

A standard approach in computational detection of regulatory elements, particular for transcription factor binding sites, is to use species similarity between orthologous promoters to influence predictions, usually by aligning promoters and then only assessing regions with sufficient similarity. The underlying hypothesis is that regions with sequence-specific function will be preferentially conserved compared to other regions. An in-built problem with this approach is that there are many other functional elements than transcription factor binding sites that give rise to preferential conservation — transcribed regions, and protein-coding regions in particular. Therefore, many tools try to anchor alignments on known TSSs. As discussed above, core promoter locations are remarkably similar between human and mouse in most cases. However, Frith *et al.*[18] showed a number of cases where a human promoter had no indication of promoter activity in the orthologous region in mouse – instead, the mouse promoter was located up- or downstream of the human promoter, and, reciprocally, had no indications of usage in human; see Fig. 14.4 for an example of this. In other words, clear examples of evolutionary turnover of transcription start sites, similarly to turnover of transcription factor binding sites, as investigated by Dermitzakis. *et al.*[17] Frith *et al.* also showed some intermediary cases, where one promoter is used prevalently in human and the other in mouse, and even cases which can be interpreted as overlapping core promoters which are diverging — similar to the concept of Kawaji *et al.* discussed above, who observed this between species, but here is it on species level. Like the study by Kawaji, *et al.* this study shows the fluidity of promoters and functional elements, and also the dangers of blindly relying on sequence conservation for determining functional elements.

14.4.4 The TSS Initiation Code and Prediction of TSS Propensity

As discussed above, data from CAGE has made most TSS predictor algorithms less attractive, due to firstly the depth of the data, secondly (and perhaps even more importantly) the insights it gave on how core promoters look; as these algorithms were largely based on models of the single TSS promoter paradigm, often requiring a TATA or Inr-like element, these are largely inadequate when trying to predict TSSs in the broader CAGE context. In fact, one of the most important questions the CAGE data brings up is the definition of a promoter, which is not as trivial as we once thought. Thus, in light of the data, initiation sites and promoter classes demand new rules and new models to capture the added complexity of the various promoter architectures. To date only one article has explored this in detail, but fortunately this study shows that the TSS initiation determinants are likely not as complex as one could fear, at least on a local level. In this study, Frith *et al.*[20] devised a hierarchical framework for the genomic organization of TSSs. For those clusters intuitively corresponding to strong core promoters, a relatively simple statistical model, just based on DNA di- or trinuclotides and their distance to TSSs, can distinguish the relative usage rate (expressed

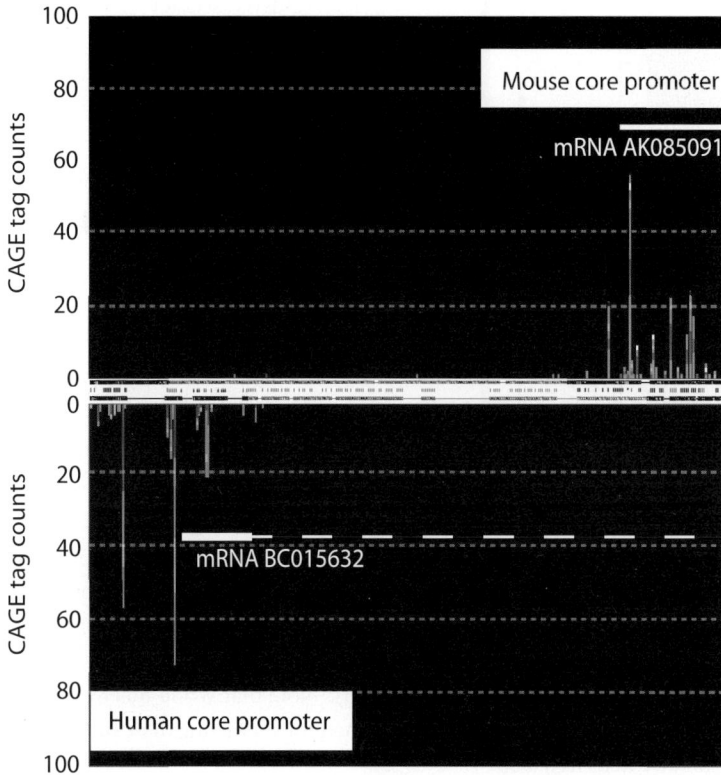

Figure 14.4. Example of evolutionary turnover of core promoters. The protein-coding regions of the AK085091 and BC015632 mRNAs from mouse and human, respectively, overlap by more than 90%. However, their 5′ start sites are not in agreement. mRNAs are in white where the thick parts are the exons; the alignment of human and mouse promoters is in the middle panel. CAGE data verifies that this is true: the location which mouse uses is devoid of CAGE tags in human, and vice versa.

as the number of CAGE tags) of individual nucleotides within the core promoter with around 90% accuracy. However, while this method can predict the TSS selection propensities within a core promoter, it cannot predict the actual *number* of CAGE tags (in other words, the actual expression rate). This strongly implies that that while the relative TSS usage in a promoter is governed by local effects, the expression strength of the whole promoter is determined by more distal, or larger-scale events, such

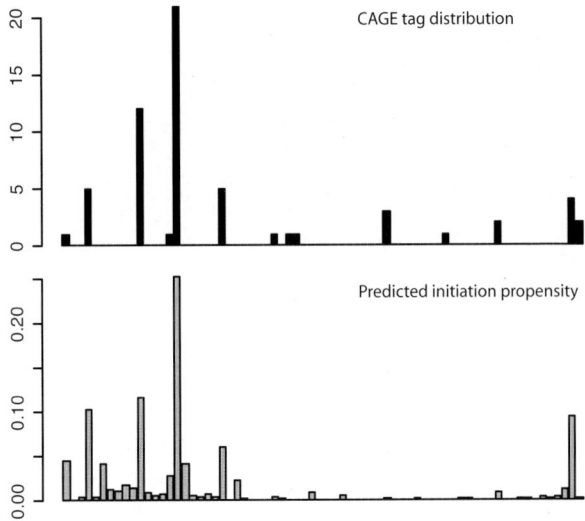

Figure 14.5. Example of predicting initiation site propensity from DNA sequence. The upper panel shows a CAGE tag cluster from HepG2 liver cells. For each nucleotide, we calculate the initiation site propensity using the local DNA sequence as a log-odds score, which is then normalized to sum to 1 over the cluster (lower panel). Note that while this example is close to ideal; this approach is not trained for predicting distributions and generally does less well.

as enhancers or histone modifications. One can view this as the role of the larger scale events are to modify accessibility and/or stabilization of a larger genomic loci where initiation takes place, while the DNA sequence within the loci will form a probability distribution describing the likely hotspots for initiation. In statistical terms, the initiation complex would then sample from this distribution, while the larger scale structures determine the rate of this sampling.

In extension, this model suggests that each nucleotide in the genome has the potential to be a TSS, although some have much lesser propensity than others and the majority will not be used significantly as they are not accessible by the transcription apparatus.

This is a widely different concept than the archetypical promoter prediction challenge, where promoters are distinct points in the genome — a "part list" reduction where it is assumed that there are sequence signals distinguishing each part from an otherwise transcriptionally silent genome. Instead, the problem is twofold — first, we need predict which regions of the genome that are accessible by the initiation machinery, which likely is mostly governed by epigenetic signals, secondly within such open regions we can apply sequence-based methods which can predict the fine-grained TSS selection. The study by Frith *et al.* can be viewed as a first stab at the second problem, while the first await pioneer studies.

ACKNOWLEDGEMENTS

This work was supported by a grant from the Novo Nordisk Foundation to the Bioinformatics Centre. We thank the FANTOM consortium and in particular Piero Carninci, Boris Lenhard, Jasmina Ponjavic, David Hume, Anders Krogh and Ole Winther for valuable comments and thoughts. Thanks to Jasmina Ponjavic for help with Fig. 14.5

References

[1] S. T. Smale and J. T. Kadonaga. The RNA polymerase II core promoter. *Annu. Rev. Biochem.* **72**, 449–479 (2003).

[2] M. Hampsey. Molecular Genetics of the RNA Polymerase II. General Transcriptional Machinery. *Microbiol. Mol. Biol. Rev.* **2**, 465–503 (1998).

[3] S. Hannenhalli and S. Levy. Promoter prediction in the human genome. *Bioinformatics* **17**(90001), S90–96 (2001).

[4] P. Carninci *et al.* The transcriptional landscape of the mammalian genome. *Science* **309**(5740), 1559–1563 (2005).

[5] P. Carninci *et al. Nat. Genet.* (2006).

[6] Y. Suzuki *et al.* Diverse transcriptional initiation revealed by fine, large–scale mapping of mRNA start sites. *EMBO Rep.* **2**(5), 388–393 (2001).

[7] M. P. Lee, K. Howcroft, A. Kotekar, H. H. Yang, K. H. Buetow and D. S. Singer. ATG deserts define a novel core promoter subclass. *Genome Res.* **15**(9), 1189–1197 (2005).

[8] M. S. Taylor, C. Kai, J. Kawai, P. Carninci, Y. Hayashizaki and C. A. Semple. Heterotachy in mammalian promoter evolution. *PLoS Genet.* **e30** (2006).

[9] S. J. Cooper, N. D. Trinklein, E. D. Anton, L. Nguyen and R. M. Myers. Comprehensive analysis of transcriptional promoter

structure and function in 1% of the human genome. *Genome Res.* **16**(1), 1–10 (2006).

[10] K. Kimura *et al.* Diversification of transcriptional modulation: Large–scale identification and characterization of putative alternative promoters of human genes. *Genome Res.* **16**(1), 55–65 (2006).

[11] F.-T. Katalin *et al.* Post-transcriptional processing generates a diversity of 5′-modified long and short RNAs. *Nature* **457**(7232), 1028–1032 (2009).

[12] S. M. Berget, Exon recognition in vertebrate splicing. *J. Biol. Chem.* **270**(6), 2411–2414 (1995).

[13] The ENCODE Consortium. Identification and analysis of functional elements in 1% of the human genome by the ENCODE pilot project. *Nature* **447**(7146), 799–816 (2007).

[14] M. G. Guenther, S. S. Levine, L. A. Boyer, R. Jaenisch and R. A. Young. A chromatin landmark and transcription initiation at most promoters in human cells. *Cell* **130**(1), 77–88 (2007).

[15] P. Kapranov *et al.* RNA maps reveal new RNA classes and a possible function for pervasive transcription. *Science* **316**(5830), 1484–1488 (2007).

[16] J. Ponjavic, B. Lenhard, C. Kai, J. Kawai, P. Carninci, Y. Hayashizaki and A. Sandelin. Transcriptional and structural impact of TATA-initiation site spacing in mammalian core promoters. *Genome Biol.* **7**(8), R78 (2006).

[17] H. Kawaji, M. C. Frith, S. Katayama, A. Sandelin, C. Kai, J. Kawai, P. Carninci and Y. Hayashizaki. Dynamic usage of transcription start sites within core promoters. *Genome Biol.* **7**(12), R118 (2006).

[18] M. C. Frith, J. Ponjavic, D. Fredman, C. Kai, J. Kawai, P. Carninci, Y. Hayshizaki and A. Sandelin. Evolutionary turnover of mammalian transcription start sites. *Genome Res.* **16**(6), 713–722 (2006).

[19] E. T. Dermitzakis and A. G. Clark. Evolution of transcription factor binding sites in mammalian gene regulatory regions: conservation and turnover. *Mol. Biol. Evol.* **19**(7), 1114–1121 (2002).

[20] M. C. Frith, E. Valen, A. Krogh, Y. Hayashizaki, P. Carninci and A. Sandelin. A code for transcription initiation in mammalian genomes. *Genome Res.* **18**(1), 1–12 (2008).

Chapter Fifteen

Future Challenges in CAGE Analysis

E. Valen* and A. Sandelin[†]

*Department of Biology & Biotech Research and Innovation Centre,
University of Copenhagen, Denmark
Email: *albin@binf.ku.dk, [†]eivind@binf.ku.dk*

Compared to other expression technologies CAGE is a relatively new approach, and as a consequence there has not been developed any standard methods with which to analyze the data. This situation is similar to the one faced when microarrays were introduced. In this chapter, we will describe first some issues worth reflecting on when analyzing CAGE data, and then describe a few particular challenges inherent in the data, which need to be addressed in future studies: the sampling depth problem, the difficulties in assessing noise and how to cluster CAGE tags in a meaningful way.

15.1 WHAT ARE WE MEASURING?

On the surface, CAGE appears to be a straightforward technology, however, for an unwary researcher there are several caveats that may lead to false conclusions. This section deals with some of the facts one needs to be aware of in statistical analysis of CAGE data.

15.1.1 Transcriptome Content — Not TSS Usage

The first thing to remark is that CAGE only indirectly measures TSS usage. The tags are sampled from the transcriptome (capped RNAs) and are therefore subject to any biological

Cap Analysis Gene Expression (CAGE): The Science of Decoding Gene Transcription **edited by P Carninci**
Copyright © 2010 by Pan Stanford Publishing Pte Ltd
www.panstanford.com
978-981-4241-34-2

processes that might affect them after transcription initiation. This includes, but is not limited to, rapid degradation, RNA silencing and antisense effects. While this is true of most methods for expression analysis it is particularly important if the goal is to measure raw TSS distributions rather than the amount of transcripts in the cell. As an example, two promoters with a different total tag counts might in fact be transcriped equally much, but a later event might be inhibiting one of the transcript products. One should also keep in mind that the influence of these processes will depend on whether tags are sampled from the cytosol, the nucleus or both, and what priming technology that is used.

15.1.2 Unknown Population Size

Hybridization-based techniques, such as microarrays, can be said to integrate over all RNAs that the probes are capable of interacting with. This is not true for sequencing-based techniques, which *sample* RNAs from a larger population. As a sampling technology, CAGE gives the user an indication of the relative usage of TSSs within the cells from an experiment. While this is well as long as the cells the user is studying are reasonably similar, it might pose problems when comparing different cells due to lack of information about the total population size of transcripts in each cell. Two different tissues, for instance, may have a widely different number of active transcripts. This implies that while questions such as whether the relative usage of a promoter is different between cells is fine, asking whether or not a promoter is used more (measured in initiation events per unit of time) can be very hard, unless the RNA content and cell count is known; note that the mass of RNA is a poor indicator as all types of RNAs cannot be captured (they have to be capped). One should not be blinded by the fact that the result of CAGE is raw counts, as these only indicate how much a TSS is used relative to other TSSs in the cell(s). Therefore it is not "absolute expression rate" in the true sense, although the units we are measuring are actual initiation events. Due to sample size effects, a single tag in one experiment might still indicate ten times higher transcription rates as in another experiment with the same sample size if the population sizes of the cells are equally disproportionate.

15.1.3 Averaging out the Differences

Finally, we would like to point out that a CAGE experiment is so far only performed on populations of cells. This means that the experiment does not necessarily reflect the transcriptome in a single cell, only the average over a large number. In some cases this may be good because it gives a better image of how the given tissue behaves, since a tissue is a collection of cells which are not fully coordinated. However, this comes at the expense that some details may be lost, notably interactions between start sites and/or promoters. A thought experiment with a pair of transcripts or promoters that in some ways are mutually exclusive nicely illustrate this — this pair might still be observed using CAGE as a collection of cells are assessed. Whether the effect is due to spatial restrictions or chromatin rearrangements or other processes the effect will not show up in a typical CAGE experiment if both promoters are used in different cells in the population. However, this is true for any technology that is applied to a cell population; as it stands now, this includes all high-throughput methods. This problem has a conceptually easy solution: performing CAGE or other analyses on a single cell, which is under development.

15.2 HOW CLOSE TO "THE TRUTH" ARE WE?

15.2.1 Sampling Depth and Saturation

While the raw numbers of tags in CAGE is impressive compared to those of individual experiments in classical molecular biology, it is clear that the typical CAGE library is but a small sampling of the RNAs in a population of cells. One easy way of realizing this is by looking at the large number of CAGE tags mapping to single positions in the genome — these are sampled only once since for mRNAs that are not widely abundant, the most likely outcome is a single hit, or none. By bootstrapping, we can also gain some understanding of the rate of saturation; that is, how large percentage of promoters or TSSs that we hit with a given sequencing depth.

In Fig. 15.1, we have sampled an increasing number of tags from 7 tissues, and the counted how many unique transcription start sites (with >=2 tags) that this will give. With no saturation, this relation would be linear: this is not the case, but one

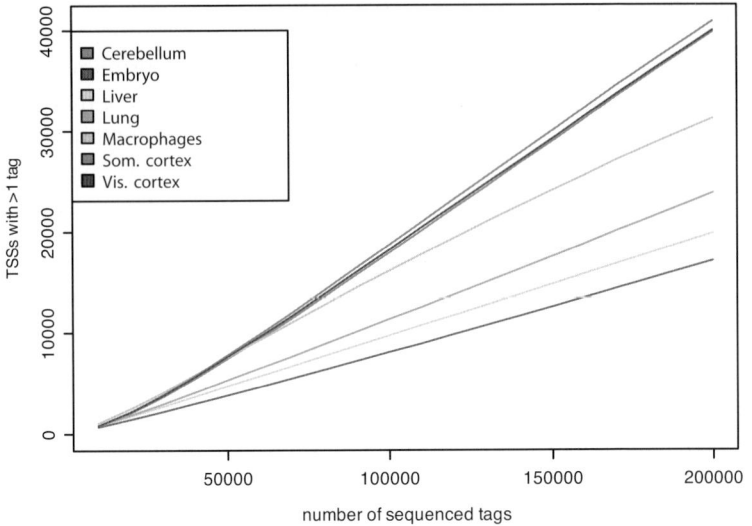

Figure 15.1. Transcription start site complexity differences in tissues. We cumulatively sample tags from a given tissue and see how many unique TSSs that are indicated with >1 tag. The Y axis values are means out of 10 different samplings.

has to sample many tags for the saturation tendencies to become apparent, which implicates that we are under-sampling. It is theoretically possible to fit a curve to these samplings and predict the plateau where the only added TSSs are noise, although there are disturbingly many parameters that are hard to estimate in any such model for any general use. For instance, the number of promoters used in a population of cells will both depend on the heterogeneity of the cells (for instance, brain tissue have a multitude of different neuron types, with different mRNA expression levels), but also with the actual complexity of the transcriptional program of a given type of cell — we have reason to believe that some cell types have many more initiation events than others. This is hard to estimate since most technologies are primed for certain types of transcripts (for instance, protein coding genes or transcripts that have no repeats).

As a simple example of modeling this behavior, we have in Fig. 15.1 without replacement randomly sampled 10,000, 20,000…etc. tags from a number of tissue libraries, and then counted the number of unique position in the genome which has at least two tags indicating that it is used as a 5' end. For simplicity, we will refer to these positions as TSSs. We repeat this analysis 10 times and report the mean number of such TSSs per sample size.

The reason for not using single tags is firstly to make it directly comparable to the analysis by Carninci *et al.*,[1] and secondly that since exact overlap of two tags is an improbable event, we can assume that those 5' ends have a low amount of false positives. If we assume the noise is low, we would expect that with sufficient number of tags, the increase in TSSs would be negligible — in other words, that we have reached a saturation of TSSs. Having an approximation of this value, and of course also the associated saturation curve, would be immediately useful, as we can then make informed decisions on how much we should sequence in a new experiment.

To start with, we sampled up to 200,000 tags from 7 mouse tissues (using data from Refs. 1 and 2) — the limit is due to that many tissues have no more tags, despite that these numbers often derived from pooled RNA libraries. An interesting issue is that the acceleration of the curves are widely different — brain tissues in particular have many more unique TSSs per tag, while embryo (note that this is a specific embryonic stage) has few TSSs/tag — implying that it, relatively speaking, uses fewer but stronger promoters. Just by looking at this picture, it is clear that the saturation limit of some tissues will be different than others, due to higher diversity of cell types as/or an intrinsically higher "TSS complexity" in cells from a certain tissue type (which is also discussed above).

As the saturation rate is low, it is hard to model this rate with these data. However, some libraries have higher amounts of tags. We will focus on liver, which is a data set of 1,218,713 tags from 12 libraries. We assume that the tag saturation follows a negative exponential:

$$Y = b(1 - e^{-aX}) \qquad (15.1)$$

where Y is the number of observed TSSs given X sequenced tags, b is a scale factor and a is a growth (or saturation) factor. Using this model, we predict the values of a and b by fitting the sampled data to the model using the nls[3] method in R[4] (Fig. 15.2).

With this assumption, we see that with the current sampling depth only encompass ∼45% of the "total" number of TSSs, and

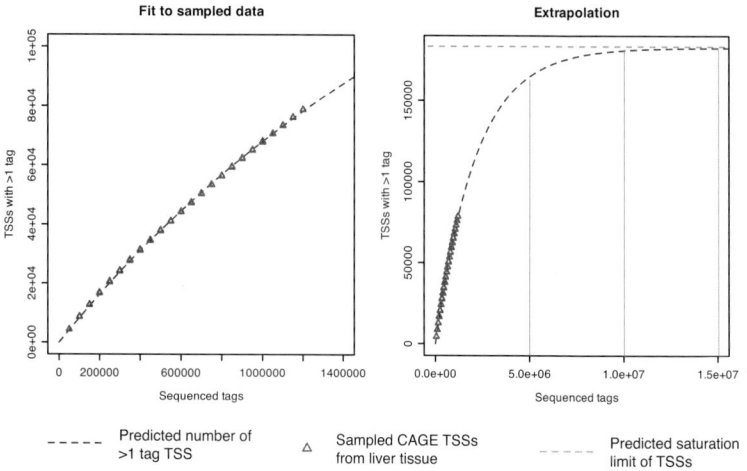

Figure 15.2. Modeling of TSS saturation in liver tissue. As in Fig. 15.1, the Y-axis shows the number of TSSs indicated by >1 tag, given the number of tags sampled (X-axis). The model is shown in Eq 1. The left panel shows the interval covered by the data, while the right panel shows an extrapolation curve, showing the predicted saturation limit. This indicates that a sample depth of 5-10 million tags would cover most TSSs.

that a sequencing depth of 5-10 million tags would be motivated as it would give a significant increase of TSSs.

An obvious extension to this simple model would be to incorporate noise: we would expect that a noise factor would always add number of TSSs that should be close to linear with the number of sequenced tags, while the number of genuine TSSs is expected to reach plateau at some stage. This in itself is a simplification since the genome size is limited, but it is a reasonable model given the size of the genome. This model can be formally stated as:

$$Y = b(1 - e^{-aX}) + kX \qquad (15.2)$$

where kX is modeling linear noise. We would then be interested in the point where the linear part of the model is the only one contributing significantly: this is where true saturation has been reached.

With sufficient data, it would be possible to assess the constants a, b, k from the data; however, with the current data we found that when using standard fitting procedures, the linear part

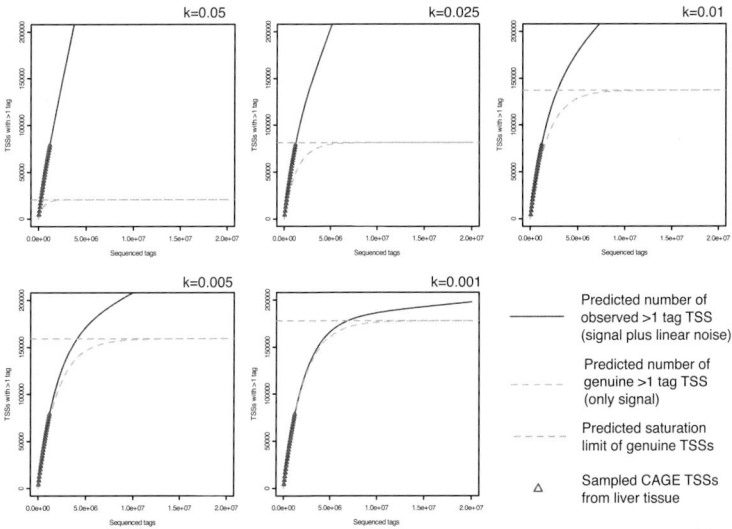

Figure 15.3. Modelling TSS saturation with a fixed noise term. The process in Fig. 15.2 is repeated, but a fixed linear noise term (k) is added to the model, which can be interpreted as the chance that a given sampled TSS on the Y-axis is due to noise (see Eq. 15.2). We tested several different noise terms, from high (0.05) to low (0.01), and show both the predicted number of TSS that one would observe when sequencing X tags, as well as the genuine number of TSSs (the noise term is subtracted). With high noise, we predict that we have seen most genuine TSSs, while if the noise is low, we are approaching the curve shown in Fig. 15.2

of the model will dominate, so that there will be no saturation tendencies. Instead, we experimented with setting the noise factor $k = 0.05, 0.025 \ldots 0.001$. This makes sense as we in fact do not have a stringent assessment of the noise level of the >1 tag TSSs that we analyze. In Fig. 15.3, we show the results as two curves: firstly, the predicted number of TSSs obtained from sequencing X tags (including noise), secondly, the same prediction with the linear noise term subtracted from the model (in other words, predicted number of genuine TSSs).

It is interesting to note the severe impact of the noise level in this model, in terms of how many genuine TSSs we expect. If the noise level is reasonably high (0.05), we have actually seen almost all genuine TSSs with the current sampling depth, along with a substantial number of noise TSSs. That is, we have reached saturation of genuine TSSs, but not the observed sampling — in fact, we will never reach a stage where saturation is measurable. Conversely, if the noise level is very low (0.001), we are approaching our previous model in Fig. 15.2, where only ~45% of the saturation of genuine TSSs has been reached (and a similar fraction of the total observed TSSs). Regardless of the noise level, with this dataset it seems like the estimate of sequence depth we made in Fig. 15.2 is reasonable: 5–10 million tags. We would like to emphasize that the results presented in this section should not be regarded hard facts — they are extrapolated from a model which is a simplification of reality, with the goal of exploration and to raise new questions. As we have discussed previously, these saturation curves will be different also due to cell heterogeneity and intrinsic transcriptional program.

However, these results indicate that that in order to adequately model the number of tags required, we also need a reasonable noise estimate. We will discuss noise and the issues on how to interpret and measure it in the next section.

15.2.2 *Methodological vs Biological Noise*

When looking at any genome-mapping and/or expression data, it is important to appreciate that a certain level of noise will always be a part of the experiments. This also applies to CAGE. For instance, there are ribosomal RNAs that are picked up by CAGE, despite that these are not supposed to be capped, which in turn is the selection criterion for CAGE.

However, it is important to consider the various types of noise. We are usually thinking about noise as a product of the experimental procedure — either random noise, such as faulty capping which is a probabilistic process or systematic noise, like the typical addition of an extra G to tags by the restriction site enzyme. However, it is our belief that a significant fraction of the low frequency tags (and also low frequency RNAs reported from tiling arrays), are "biological noise" — spurious transcription that give no added fitness but is a byproduct of the machinery the cell is operating (also discussed in Ref. 5). We should bear in mind that biological enzymes and connected processes are subject to the same chemical laws as other reactions: events are described by rates, which in turn can be viewed as probabilistic processes. This is also true for transcription, which should be regarded as a probabilistic process. This means that the deterministic "lock-and-key" line of thinking suggested in many textbooks is not particularly realistic.

It is hard to think about ways to avoid biological noise, except by filters such as RNA size and maturation stages of RNAs. It is also not certain that we want to avoid it, since we are measuring biological processes.

Going back to the issue with technical noise, we can be more confident about TSS locations and their relative strength by the use of technical replicates. Of course, since we are under-sampling the total number of RNAs, we cannot expect a perfect correlation between replicates, especially if the tag count is low.

In terms of location assessment ("is this position that we indicate by CAGE a true TSS"), it is hard to rigorously assess the level of noise, as this will be dependent on other validation techniques as (i) these will not have perfect sensitivity or specificity, (ii) we would have to assess a substantial number of TSSs, which is hard to do practically. However, as this noise level makes a huge impact in terms of how much we need to sequence, this is an important challenge for future studies.

15.2.3 Clustering CAGE Tags in Meaningful Way

The goal of clustering is ultimately to identify the region or group of tags subject to the same transcriptional regulation inputs and by extension the same controlling machinery. The clusters lay the foundation for much of the downstream analysis protocols and are consequently imperative to most CAGE experiments. Decisions made during clustering will as a consequence influence the results of other analyses and it is therefore necessary to consider these choices carefully.

Since the biological processes regulating transcription initiation is not fully known it is hard, if not impossible, to partition the tags adequately according to purely biological rules. Most methods have therefore concentrated on inferring the relationship between tags indirectly through statistical arguments, using information such as proximity and/or co-regulation.

In CAGE context the basal cluster is usually referred to as a tag cluster and is an extension of the classical core promoter to a framework with multiple TSSs. The initial method proposed by Carninci *et al.*[1] accomplished this by simply grouping all overlapping tags on the strand. This, while justified in that we then know for certain that the transcripts of the tags overlap, unfortunately rests on the length of a tag (~20 bp), which has no biological motivation. There are also some practical problems with this approach to consider:

For most studies, clusters having few tags are not as interesting as those with many tags (as described above); since there are many singleton tags, most clusters will only contain one tag. This means that when using the clusters for analysis, an arbitrary cutoff (the number of tags inside the clusters) had to be introduced.

There are quite a few cases where single tags that have unique starts tile a larger region. As most studies want to consider denser aggregates of tags, this is not optimal. We can also ponder what would happen if we in the future had an extremely deep sampling — would we in fact only have one cluster per chromosome and strand?

Conversely, we can consider the opposite problem where two clusters stand out by eye but are separated by a mere 21 bp, and thus are considered separate.

At least two approaches has since been suggested that do not rely on such *ad hoc* numbers, and also add biologically motivated new features. We will doubtless see more refined approaches in the future, but it is instructive to compare these two methods as they follow different principles, and have different strengths. Erik van Nimwegen *et al.*[8] have devised a system based on proximity and scalability of tags, where it is possible to build in certainty of TSS positions using replicates. In-built in this concept is a noise-call, which essentially makes the decision if a certain aggregate of tags are believed to be real. It also introduced the notion of looking at clusters on at least three different levels — individual nucleotide initiation events that are significant, and then different aggregates of those.

Frith *et al.*[6] introduced the concept of maximal scoring segments to TSS analysis, which is radically different from the van Nimwegen approach (although with reasonable parameters, the TSS clusters produced by the two methods are similar; in fact, part of the development of these two cluster methods was done in parallel). A maximal scoring segment (MSS) is informally a segment with more than x tags per nucleotide where the removal of any suffix or prefix will results in a lower mean.[7] This gives a hierarchical clustering of tags at various means and also enables us to identify the highest and lowest x where a segment is an MSS.

A segment which is an MSS over a large range of values for x is intuitively a stable cluster since the distribution of tags must change extensively for the cluster to break up. To capture this notion we introduced a stability measure of a cluster defined as $\max x / \min x$. The beauty of this method is that it uses a minimum of assumptions about the data, relying mainly on the distance and strength of the TSSs — and it can conceptually be used on any data that can be mapped to genomes.

A difference to the method above is that while both produce different levels of clusters, the MSS method can conceptually give an infinite number of levels; conversely, it has no in-built noise cutoff. While the clusters-within-clusters data may be harder to handle in analysis, it may be closer to the biological reality. As there are no bounds on the sizes of cluster, meaning that it will capture such events as exonic TSSs scattered over one gene as a weak cluster that stands out from the background distribution of tags. This means that these cluster may indicate events on larger scale — in extension, one can view differently sized clusters as being produced by processes working with different resolution. This hints on the possibility of using CAGE data to indirectly probe events on epigenetic/chromosomal scales.

References

[1] P. Carninci *et al. Nat. Genet.* (2006).

[2] P. Carninci *et al.* The transcriptional landscape of the mammalian genome. *Science* **309**(5740), 1559–1563 (2005).

[3] D. M. Bates and D. G. Watts. *Nonlinear Regression Analysis and Its Applications*, Wiley (1998).

[4] R. Ihaka and R. Gentleman. R: A language for data analysis and graphics. *J. Comp. Graph. Stat.* **5**(3), 299–314 (1996).

[5] The ENCODE Consortium. Identification and analysis of functional elements in 1

[6] M. C. Frith, E. Valen, A. Krogh, Y. Hayashizaki, P. Carninci and A. Sandelin. A code for transcription initiation in mammalian genomes. *Genome Res.* **18**(1), 1–12 (2008).

[7] W. L. Ruzzo and M. Tompa. A linear time algorithm for finding all maximal scoring subsequences. *Proc. Int. Conf. Intell. Syst. Mol. Biol.* 234–241 (1999).

[8] FANTOM Consortium. The transcriptional network that controls growth arrest and differentiation in a human myeloid leukemia cell line. *Nat. Genet.* **41**(5), 553-562 (2009).

Chapter Sixteen

Comparative Genomics and Mammalian Promoter Evolution

Martin S. Taylor[1,*], Gregory Jordan[1] and
Colin A. Semple[2]

[1]*European Bioinformatics Institute, UK*
[2]*MRC Human Genetics Unit, Western General Hospital, UK*
*Email: *mst@ebi.ac.uk*

Although there is general agreement on the importance of regulatory changes in molecular evolution we are still in the early stages
of examining the evolution of the genomic regions that are most
rich in regulatory information: promoters. This review outlines
strategies to combine CAGE data with other datasets to perform
meta-analyses of mammalian promoter evolution. We consider the
advantages of CAGE data for the identification of promoter regions
and demonstrate how these regions can be used to investigate the
spatial distribution of constraints within promoters as well as differences in the pattern of constraint between promoter functional
classes (CpG island containing, TATA box containing etc). Comparative genomics together with CAGE derived promoter regions
have also allowed important insights into the complex evolutionary dynamics of promoters across mammalian orders. All measurable rates of evolutionary divergence have been shown to vary,
both between functional classes of promoters and between different lineages. The most dramatic differences are found between
the primate lineage and other non-primate mammals, with a pronounced acceleration in primates now reported by several groups.
The causes of such differences remain unknown, but positive selection of favourable variants, elevated mutation rates and disrupted
purifying selection mediated by low population sizes have all been
suggested as possible explanations. Comparative genomics alone
cannot determine the relative importance of each of these causes,

*Cap Analysis Gene Expression (CAGE): The Science of Decoding Gene
Transcription* **edited by P Carninci**
Copyright © 2010 by Pan Stanford Publishing Pte Ltd
www.panstanford.com
978-981-4241-34-2

however recent additional datasets may allow us to address this question.

16.1 INTRODUCTION

The enzyme complex RNA polymerase II (RNApolII) is responsible for the production of the majority of diverse transcripts in eukaryotic cells, including messenger RNAs (mRNA) that encode proteins.[37] The loading of RNApolII onto DNA, its positioning and ultimately the location of the transcription start site (TSS) are traditionally thought to be guided by sequences within the core promoter regions, located immediately upstream of the TSS.[21,26] The exact definition and extent of a core promoter varies between studies, but for the purposes of our discussions we have defined the core promoter as the 150 to 200 nucleotides upstream of the TSS on which the RNApolII preinitiation complex assembles. Most, and perhaps all, RNApolII core promoters have the ability to drive a basal level of transcription,[40] but that level of transcription is extensively modulated through the action of *cis*-regulatory elements which can be found hundreds of kilobases from the promoter they modulate (e.g. Ref. 23).

Despite the long range over which particular *cis*-regulatory elements can act, recent large-scale studies have been remarkably consistent in finding that the functional sites regulating the activity of promoters are concentrated within a few thousand bases of the TSS. One major study defined both TSSs and regions enriched for the binding of known regulatory proteins using high-throughput experimental methods and found that regulatory protein binding sites congregate in 'regulatory clusters'. Two thirds of these clusters were found within a 2.5 Kb radius either side of a TSS.[8] Similarly, a large study seeking SNPs with regulatory effects on the expression of neighbouring genes found that most variants exercising a significant effect were located within 2 Kb of a TSS.[38] The concentration of functional sites in the locality of the TSS suggests that most of the regulatory code orchestrating gene expression is concentrated in these confined regions. As such they represent a rich vein of the genome in which to identify phenotype and disease associated polymorphisms, and investigate the regulatory circuits of the genome. They also allow us to explore the genetic adaptations between species that have long been anticipated to involve regulatory mutations more often than changes to protein coding sequence.[43] For these reasons there is a great deal of interest in understanding the patterns and rates of sequence evolution in the sequences flanking TSSs.

As regions containing significant densities of functionally important sites, it would generally be expected that TSS flanking regions would be subject to selective constraint and typically exhibit a slower rate of sequence change (nucleotide substitution, insertion and deletion) than non-functional sequences in the genome. This expectation appeared to be validated by early studies of mammalian promoter evolution, comparing modest numbers (n=77, n=100) of human and mouse orthologous promoters and intergenic regions,[14,35] which found evidence of evolutionary constraint in regions upstream of the presumed TSS. However, since rodents and primates diverged perhaps 90 million years ago it remained possible that different patterns would be observed within different mammalian lineages. Keightley and Gaffney[16] examined substitution rates at 300 loci orthologous between mouse and rat, and showed significant selective constraints across promoter regions in the rodent lineage. In a later study the same group compared the divergence seen between 300 mouse and rat promoters with that seen between 1000 human and chimpanzee promoters with surprising results: promoters appeared to be evolving effectively neutrally within the primate lineage and lacked any detectable selective constraint.[17] In fact primate promoter regions appeared to be evolving somewhat faster than the intronic regions used as near-neutral controls (see section below on neutral rate estimates), though this trend did not reach statistical significance. In a similar study, Bush and Lahn[2] examined promoter regions for 5547 human-chimpanzee orthologous gene pairs and also detected less constraint in primates than in rodents across entire promoters, but they also showed that selective constraint does operate upon small (16 bp) promoter sub-regions. These sub-regions correspond to sites conserved across mammalian or vertebrate species. These two phenomena, accelerated divergence in primate promoters on the one hand and conserved promoter sub-sequences on the other, can be reconciled by the proposition that promoters have a low density of short, functional sites embedded in longer neutrally evolving regions.

In this chapter we show how CAGE can be combined with other genomic resources, in particular cross-species alignments and polymorphism data, to advance the studies described above, investigating the general trends in mammalian promoter evolution. In doing so we illustrate the considerable advantages, and also the limitations, of CAGE data for the definition of TSS and promoter regions. Subsequent sections highlight recent insights into gene regulation and evolutionary processes that have come largely from CAGE based evolutionary analysis. We also demonstrate how high quality and precise TSS definitions can be used to investigate the spatial distribution of constraints in TSS flanking regions and differences in the pattern of constraint between promoter categories. We then discuss the questions that have been raised by these findings, and consider how they in turn can be addressed.

16.2 RESOURCES FOR COMPARATIVE GENOMIC ANALYSIS

16.2.1 *Precise Definition of Transcription Start Sites by CAGE*

Prior to the availability of CAGE data for the definition of TSSs, studies of promoter evolution have used translational start sites as a surrogate for the TSS (e.g. Ref. 17), or they have based the TSS definition on mRNA sequence data, or more recently RefSeq or Ensembl annotation of the 5′ end of the transcript.[12] An obvious limitation of the translational start site as a surrogate for the TSS, is that many transcripts include 5′ non-coding exons.[5] In such cases the promoter is likely to be missed entirely in the analysis. While full-length mRNA sequences and their curated cousins: the RefSeqs, often do capture non-coding upstream exons and much of the 5′ UTR, RACE and CAGE have demonstrated that there is considerable biological variability in the precise location of transcription initiation.[5] Figure 16.1 shows that the great depth of CAGE data relative to mRNA and EST coverage allows that biological variability to be measured for a given TSS, and a consensus position (typically the modal tag) to be defined. This provides a considerable advantage over other annotations, in the definition of orthologous promoter regions between species (Fig. 16.1), and in the investigation of spatial constraints relative to the TSS.[39] It is worth

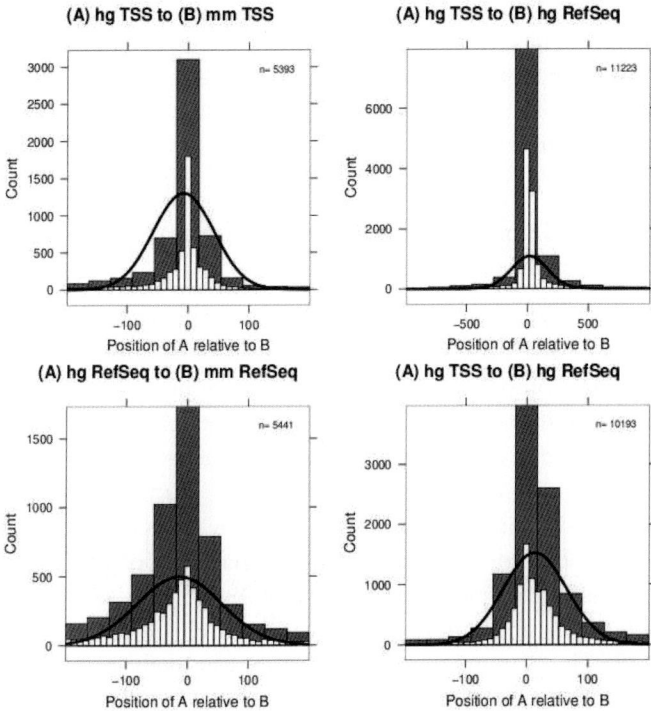

Figure 16.1. Histograms illustrating the spatial separation (nucleotides, x-axis) between TSS mapping methods. CAGE defined TSS are derived from Carninci *et al.*.[5] after filtering for >2 tags per million averaged over all CAGE libraries that contain >30,000 tags. Blue columns represent the same data as white columns but at larger bin sizes (36.4 bp and 9.8 bp respectively), and the variable n indicates the number of data points in the current view. The top left graph shows the distance between reference tag positions of orthologous promoters when mapped between human (hg) and mouse (mm) reference genomes. Where the same mapping is performed between annotated human (hg) and mouse (mm) RefSeq 5′ ends there is greater variation (bottom left graph). Rightmost panels show the distribution of separation between CAGE defined TSS reference positions and annotated RefSeq 5′ ends within the human genome. The lower right graph shows a magnified view of the data depicted in the upper right graph (blue and white columns represent 181.8 bp and 48.8 bp respectively for this graph only). It is evident from these plots that RefSeq annotation tends to exaggerate the length of 5′ UTR sequences (right hand skew to the distribution), apparently through the selection of the most 5′ extending piece of transcript evidence rather than the consensus 5′ end indicated by transcript data (unpublished observations).

noting that until recently the extent and complexity of mammalian transcription has been hugely underestimated such that the available data in commonly used genome annotation databases omits the majority of TSSs and therefore promoters.[4,5,8] The CAGE data is able to capture much of that complexity.

16.2.2 CAGE as a Quantitative Measure of Expression

Studies of promoter evolution are often concerned with the expression profile driven by the promoter, for example in the comparison of evolutionary constraints between housekeeping genes and those with more restricted expression.[45] CAGE data is intrinsically quantitative, with the frequency of tags corresponding to the relative level of expression. In contrast, studies using RefSeq 5' end and similar annotations of TSSs typically have to base their measures of promoter activity on the mapping of promoters to independent datasets of microarray probes.[45] This assignment is often tenuous and ambiguous given the high frequency of alternate TSSs in mammalian genomes,[5] and is likely to be a major contributor to the poor correlation observed between expression measured by CAGE tag clusters and microarray hybridisation to downstream probe sets.[9]

16.2.3 Alignments for Comparative Genomics

The starting point for the majority of comparative genomic analyses, including those focussing on non-coding regions such as promoters, is an alignment between homologous sequences. There are many tools available to construct pairwise and multiple sequence genomic alignments (reviewed by Taylor and Copley, 2007), but there are many potential pitfalls to the reliable identification of orthologous sequences and the production of high quality multiple sequence alignments. For this reason and for ease of reproducibility, it often makes sense to utilise publicly available genomic alignments from central resources such as the UCSC Genome Browser (http://genome.ucsc.edu) or Ensembl (http://www.ensembl.org). The UCSC LiftOver tool is widely used to define orthologous sites or regions between genomes. Ensembl has recently produced true multiple sequence alignments of whole mammalian genomes. This is a substantially improved resource over the stacking of pairwise alignments relative to a reference genome, and allows aligned regions of interest to be directly imported into computational pipelines for phylogenetic analysis.

16.2.4 Identifying Spatial Constraints and Patterns

Although each TSS is likely to have its own specific set of cis-regulatory elements, we have already discussed that they generally seem to be concentrated into the TSS flanking regions. By averaging across many promoter regions Taylor et al.[39] were able to investigate the spatial distribution of constraints around TSSs down to single nucleotide resolution (Fig. 16.2). This revealed an asymmetric distribution of constraint around TSSs, with a high level of constraint immediately upstream of the TSS. This high level of constraint reduced rapidly within 200 nucleotides of the TSS but still showed significant constraint out to at least 1 Kb upstream. The high resolution of this study allowed the identification of periodic patterns of conservation in the core promoter, at least some of which correspond to known signals such as the consistent spacing of the TATA box relative to the TSS.[33]

Comparison of substitution rate profiles for different promoter categories was particularly revealing. For instance Taylor et al.[39] were able to show that promoters overlapping CpG islands have higher selective constraint in close proximity to the TSS than non-CpG island promoters, but also that there is more constraint further upstream in the non-CpG island promoters. The may reflect the more pronounced difference in upstream substitution rates observed between TSS defined by high average levels of expression versus low average levels of expression (Fig. 16.2). This result could indicate that housekeeping promoters (those expressed similarly in all tissues and conditions) have most of their cis-regulatory constraints confined to the core promoter region, whereas genes that are more dynamically regulated, dependent on tissue, developmental stage or external stimulus have higher densities of more distant cis-regulatory sequences. Unfortunately this study did not discriminate between broad expression across many samples and high levels of expression restricted to particular tissues or conditions. However, this hypothesis has recently been supported by the work of Farre et al. (2007)[45] who found regions upstream of housekeeping genes have fewer alignable nucleotides between human and mice than genes with more restricted patterns of expression.

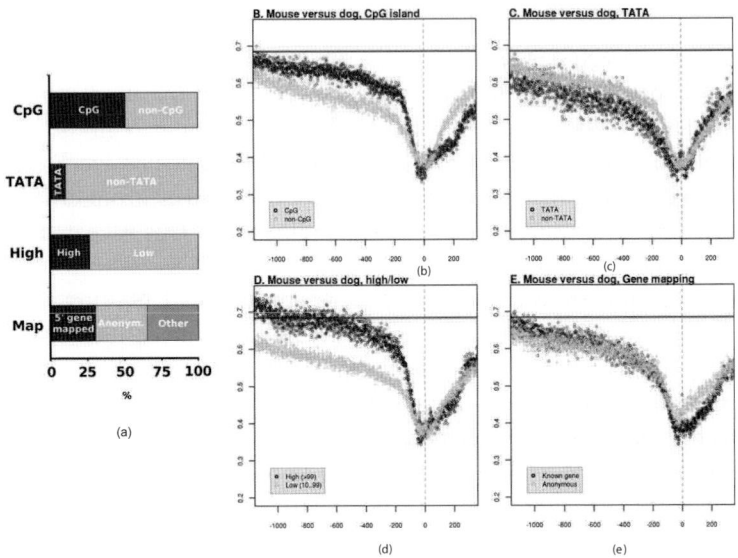

Figure 16.2. (a) The percentage of all mouse TSSs assigned to each category. Dark blue shows the percentage assigned to the category annotated to the left, and light blue the reciprocal category (e.g. non-CpG is the reciprocal of CpG). The colour coding is consistent with (b-e). The "Map" category refers to whether the TSS could be mapped to the annotated 5'-most end of a known protein-coding gene (dark blue), could not mapped to a gene (light blue), or maps internally to an annotated gene structure (grey). (b-e) Single nucleotide resolution estimates of substitution rates (K, substitutions per aligned nucleotide) calculated from promoters assigned to the indicated categories. Only rates calculated from mouse-dog comparisons are shown. The 95% confidence intervals have been excluded for clarity. Red horizontal lines show K for ancient repeat sequences, nucleotide position is shown on the x-axis relative to the TSS at +1 (grey vertical line), and K is shown on the y-axis. Although there are three categories indicated for gene mapping in (a), only two are shown for clarity. Figure reproduced from Taylor *et al.*[39] under the Creative Commons Attribution License.

16.3 GENOME WIDE TRENDS IN MAMMALIAN PROMOTER EVOLUTION

The study of spatial constraint we have considered above concentrated on relative measures of constraint, whether a particular nucleotide position has a higher or lower substitution rate than other sites in the promoter region. The study of Taylor *et al.*[39] also used an external reference dataset of sequences assumed to be non-functional, to provide an estimate of the neutral evolutionary rate (Fig. 16.2). This represents an important concept in comparative genomics that we briefly explain as it has important implications for discussions in this and subsequent sections. The underlying principal of comparative genomics is to use the signal of past selection as an assay for biological function in genomic sequence. Natural selection can only act on genetic variation that manifests itself as phenotypic differences between individual organisms of a population. It is a stringent filter: even a 0.001% reduction in reproductive success will lead to a polymorphism being reliably removed from most mammalian populations.[29] In the absence of selection, sequences are said to evolve neutrally and, given some assumptions, the neutral rate is approximated by the mutation rate.[19] Any significant deviation from the neutral rate is diagnostic for the action of selection. It follows that the neutral rate estimate is crucial for comparative genomics studies and an inappropriate estimate could lead one to wrongly infer the action of selection.

As mentioned above, Keightley *et al.*[17] reported that divergence in primate promoters was not significantly different from that seen in a putatively neutral dataset of intronic sequences, and concluded that there had been a catastrophic loss of constraint in primate promoters. They have argued that this is due to the fixation of slightly deleterious mutations in primates that are selectively eliminated in larger rodent populations.[18] Bush and Lahn[2] have also emphasised the roles that positive selection and mutational bias could play in accelerated promoter divergence. In contrast, Taylor *et al.*[39] found that with a larger dataset based upon experimentally defined TSSs, primate promoter divergence consistently and significantly exceeds that seen in near-neutral controls. It is important to note that this degree of accelerated evolution is not explicable by a relaxation of selective constraint alone (where we would expect no significant deviation from neutrality) and suggests additional forces have been in play. Candidates include increased positive selection and elevated mutation rates at primate promoters, but the extent to which one or more apply is unknown. It is also worth noting that substitution rate measures by themselves do not allow us to confidently unravel the extent to which positive selection, relaxed constraint or elevated mutation rates have contributed to accelerated substitution rates across a population of promoters. This was essentially the conclusion of a recent study of 2492 core promoters in primates.[29] The authors detected a significant elevation of substitution rates relative to near-neutral control sequences across all promoters and, like Taylor *et al.*,[39] identified positive selection or elevated mutation rates as the possible explanations. They examined 24 of the most rapidly evolving core promoters as the most likely to be targets of positive selection and found some corroborating evidence of this using polymorphism data.[24]

16.4 PROMOTERS REPRESENT AN UNUSUAL GENOMIC ENVIRONMENT

It has long been known that mammalian promoters represent an unusual environment relative to the bulk genome, for example in their sequence composition, chromatin state and methylation status. Their distinctions from the rest of the genome are of great biological interest, but also generate problems for their analysis. This is particularly true of evolutionary studies where, it is often desirable to make direct comparisons of the patterns and rates of sequence change between promoter and non-promoter sequences. Below we highlight three aspects of promoter biology that could have considerable impact on the evolutionary behaviour of promoters.

16.4.1 Sequence Composition

Compositional bias in promoter sequences is a familiar problem for those undertaking regulatory site prediction or evolutionary studies. Most commonly this is encountered as a high frequency of GC rich regions, usually associated with CpG islands, and a number of methods have evolved to take account of such phenomena. Such methods range from masking (i.e. omitting) predicted

CpG islands to masking any nucleotide that is currently or is likely to have been a member of a CpG dinucleotide (e.g. Taylor *et al.*)[39] These issues are particularly important in comparisons between mammalian lineages, given the well documented differences in CpG-island architecture between them, especially between rodent and primate genomes.[41] However, it is becoming clear that there are more subtle, though equally pervasive patterns of compositional bias present within promoter sequences. Frith *et al.*[10] have shown that dinucleotide frequencies within core promoters are highly nonrandom, and have argued that these frequencies constitute a code for TSS selection. It seems that RNA Polymerase II and associated factors scan accessible DNA, and transcription is more likely to be initiated within sequence contexts that fit a particular distribution.[10]

16.4.2 *Chromatin Structure at Promoters*

Although we are familiar with the human genome sequence as a collection of linear DNA sequences, it actually exists within cells as an intricately structured environment organised in several hierarchical layers. Each layer is based on interactions between DNA and proteins to achieve compaction of the linear DNA helix, and the term "chromatin structure" covers all of these layers up to the chromosome. The eukaryotic nucleosome core particle is formed by a 147 bp section of DNA wound around an octomer of eight histones. The nucleosome is the critical building block of chromosome structure and the folding of long nucleosome arrays forms the basis for all higher order, secondary chromatin structures.[42] For instance, individual nucleosomes are joined to each other by a linker histone (H1) and an intervening stretch of DNA to yield the 10 nm chromatin fiber, and this may be further compacted into the helical 30 nm fibre via interactions between histone tails. Chromatin structure is known to affect evolution at a number of levels from overall chromosome architecture to local mutation rates[31,34] and may therefore be relevant to the acceleration of substitution rates seen in primate promoters.

Recent genome-wide studies of chromatin structure have revealed the special landscapes that promoters occupy by mapping Dnase I hypersensitive (Dnase I HS) sites. It has been found that both promoters and more distal (presumably long range) regulatory elements have relatively decondensed, "open" chromatin structures and among these accessible regions the promoters of protein coding genes are particularly "open" (i.e. extremely hypersensitive to DNase I digestion). It is thought that these open regions are nucleosome depleted to allow access to the transcriptional and regulatory machinery. Although most Dnase I HS sites in the human genome do not show any evidence for selective constraint, most of those located at promoters are detectably conserved to some degree across a variety of non-primate mammals.[1,8] Furthermore the pattern of nucleosome depletion (identified as regions with a paucity of histone marks) shows a marked correspondence with the pattern of constraint observed at mammalian promoters.[39] Specifically, the most conserved region within 200 bp of the TSS corresponds to a precipitous drop in the frequency of various histone modifications. Also, genes that are both highly expressed and are enriched for housekeeping functions (such as mRNA processing and splicing and general metabolic processes) have an extremely hypersensitive promoter region,[1] and as already mentioned, these promoters are among the most divergent in the mammalian genome.[39] These observations suggest that there is a good correspondence between the chromatin structure and substitution rates observed at promoters.

16.4.3 *Promoters and Meiotic Recombination*

There is a well established relationship between hotspots of meiotic recombination and promoters in the yeast genome.[28] Recombination hotspots have been identified in the human and mouse genomes[27,36] and show a subtle but significant enrichment in promoter regions.[13] This is of particular interest as recombination itself may be mutagenic,[22] has been associated with bursts of lineage specific accelerated evolution[7,30] and is positively correlated with regional fluctuations in substitution rate and G+C content across mammalian genomes.[11] It is tempting to speculate that an interrelationship between promoter activity and the initiation of recombination might help explain both the unusual sequence composition and apparently higher than the genome average substitution, insertion and deletion rates in primate promoters.

16.5 INTEGRATION OF POPULATION GENETIC DATA WITH COMPARATIVE GENOMICS

The theoretical basis underlying this combined analysis can be briefly summarised: the majority of sequence changes observed between species are likely to be fixed (frequency of 1.0;).[20] Even very-mildly deleterious alleles are unlikely to rise to a high frequency or become fixed in a population. In contrast, polymorphisms sampled from within a species have not yet run the full gauntlet of selection and deleterious alleles are likely to be present. The process of mutation will have contributed equally to inter-species variation and polymorphism but selection will have had a greater influence on inter-species variation. This difference in expectation can be exploited to separate signals of selection from mutation e.g. Ref. 25. Unfortunately such combined analysis approaches are extremely sensitive to acquisition bias and the majority of intra-species variation data currently available in mammals has been subject to poorly defined ascertainment biases that cannot readily be corrected.[6] This makes the use of polymorphism data in genome-wide studies challenging. This situation is about to change with the launch of several human genome re-sequencing projects, including the 1000-genome project (http://www.1000genomes.org/). Population genetics offers many tests of polymorphism data that can identify the signature of positive selection, but in the best case scenario these data should also be supported by functional studies to directly measure the consequences of any putatively selected variant. One exemplar study examined the *cis*-regulatory evolution of the opioid neuropeptide precursor prodynorphin (PDYN) in primates.[32] Phylogenetic comparisons among several primate genomes uncovered a 68 bp regulatory element 1250 bp upstream of the TSS of the PDYN gene undergoing accelerated substitution rates in the human lineage. The acceleration was shown to be specific to this small part of the promoter and to the human lineage alone, making positive selection a likely explanation. However the authors obtained supporting evidence from two other lines of evidence. Firstly, they examined human SNP data in the vicinity of the 68 bp element and found high derived allele frequencies. This is expected as a by-product of positive selection since selected mutations often drag closely linked to alleles to high frequency. No such effect was found in chimpanzee polymorphism data. Furthermore the population genetic data was consistent with positive selection driving differentiation between current human populations while decreasing variation within populations. Secondly, they examined the effect of the 68 bp element on PYDN expression using *in vitro* assays. They were able to demonstrate that the human element drives significantly higher expression than the equivalent chimpanzee sequence regardless of flanking sequence. Together these different lines of independent evidence suggest a scenario where positive selection has helped to shape the expression pattern of a human gene by favouring promoter variants with a measurable effect on transcription.[32]

The importance of using multiple lines of evidence to infer positive selection in non-coding sequences was underlined by a recent study of 6,280 primate promoter regions.[12] As with the other studies mentioned already these authors set out to identify promoters with an accelerated substitution rate relative to near-neutral control sequences, and identified 575 with the greatest acceleration as targets of positive selection. No corroborating evidence was sought from population genetic tests and poor agreement was found with a dataset measuring expression divergence between human and chimpanzee. In addition it would appear that the set of positively selected promoters identified by Haygood *et al.*[12] overlaps substantially with those identified as being likely candidates for elevated mutation rates[47]

16.6 CONCLUDING REMARKS

There is still much to learn about mammalian promoter evolution, and the current availability of empirically determined (CAGE-derived) promoter locations combined with whole-genome multiple sequence alignments offer opportunities to answer fundamental questions. Across mammalian lineages we still lack information on the causes of evolutionary rate variation between different promoter classes, but recent data may allow us to assess the impact of differences in chromatin structure. The causes of accelerated substitution rates at primate promoters are among the most pressing among current research questions in mammalian promoter evolution. Although positive selection, elevated mutation rates and relaxation of purifying selection may all have played roles, their relative contributions are unknown. However there may be grounds for optimism.

The expanding empirical data on transcription factor binding sites within promoters may offer important insights into the evolutionary behaviour of functional sites. Also the use of human polymorphism data undoubtedly represents a good opportunity to shed light on this question, and through the reliable identification of specific positively selected promoter variants may give a new view of the past and present trajectory of human genome evolution.

References

[1] A. P. Boyle, S. Davis, S. Shulha, P. Meltzer, E. H. Margulies, Z. Weng, T. S. Furey and G. E. Crawford. High-resolution mapping and characterization of open chromatin across the genome. *Cell* **132**, 311–322 (2008).

[2] E. C. Bush and B. T. Lahn. Selective constraint on noncoding regions of hominid genomes. *PLoS Comput Biol.* **1**, e73 (2005).

[3] E. C. Bush and B. T. Lahn. Authors' reply. *PLoS Comput Biol.* **2**, e26 (2006).

[4] P. Carninci, T. Kasukawa, S. Katayama, J. Gough, M. C. Frith. *et al.* The transcriptional landscape of the mammalian genome. *Science* **309**, 1559–1563 (2005).

[5] P. Carninci, A. Sandelin, B. Lenhard, S. Katayama, K. Shimokawa, J. Ponjavic, C. A. Semple, M.S. Taylor, P. G. Engstrom and M. C. Frith *et al.* Genome-wide analysis of mammalian promoter architecture and evolution. *Nat. Genet.* **38**, 626–635 (2006).

[6] A. G. Clark, M. J. Hubisz, C. D. Bustamante, S. H. Williamson and R. Nielsen Ascertainment bias in studies of human genome-wide polymorphism. *Genome Res.* **15**, 1496–1502 (2005).

[7] T. R. Dreszer, G. D. Wall, D. Haussler and K. S. Pollard. Biased clustered substitutions in the human genome: The footprints of male-driven biased gene conversion. *Genome Res.* **17**, 1420–1430 (2007).

[8] ENCODE Project Consortium. Identification and analysis of functional elements in 1% of the human genome by the ENCODE pilot project. **447**, 799–816 (2007).

[9] G. J. Faulkner, A. R. Forrest, A. M. Chalk, K. Schroder, Y. Hayashizaki, P. Carninci, D. A. Hume and S. M. Grimmond. A rescue strategy for multimapping short sequence tags refines surveys of transcriptional activity by CAGE. *Genomics* **91**, 281–288 (2008).

[10] M. C. Frith, E. Valen, A. Krogh, Y. Hayashizaki, P. Carninci, A. Sandelin. A code for transcription initiation in mammalian genomes. *Genome Res.* **18**, 1–12 (2008).

[11] R. C. Hardison, K. M. Roskin, S. Yang, M. Diekhans, W. J. Kent, R. Weber, L. Elnitski, J. Li, M. O'Connor, D. Kolbe, S. Schwartz, T. S.

Furey, S. Whelan, N. Goldman, A. Smit, W. Miller, F. Chiaromonte and D. Haussler. Covariation in frequencies of substitution, deletion, transposition, and recombination during eutherian evolution. *Genome Res.* **13**, 13–26 (2003).

[12] R. Haygood, O. Fedrigo, B. Hanson, K. D. Yokoyama and G. A. Wray. Promoter regions of many neural- and nutrition-related genes have experienced positive selection during human evolution. *Nat. Genet.* **39**, 1140–114 (2007).

[13] International HapMap Consortium. A second generation human haplotype map of over 3.1 million SNPs. *Nature* **449**, 851–856 (2007).

[14] N. Jareborg, E. Birney and R. Durbin. Comparative analysis of non-coding regions of 77 orthologous mouse and human gene pairs. *Genome Res.* **9**, 815–824 (1999).

[15] H. Kawaji, M. C. Frith, S. Katayama, A. Sandelin, C. Kai, J. Kawai, P. Carninci and Y. Hayashizaki Dynamic usage of transcription start sites within core promoters. *Genome Biol.* **7**, R118 (2006).

[16] P. D. Keightley and D. J. Gaffney. Functional constraints and frequency of deleterious mutations in noncoding DNA of rodents. *Proc. Natl. Acad. Sci. USA* **100**, 13402–13406 (2003).

[17] P. D. Keightley and M. J. Lercher, A. Eyre-Walker Evidence for widespread degradation of gene control regions in hominid genomes. *PLoS Biol.* **3**, e42 (2005).

[18] P. D. Keightley, M. J. Lercher and A. Eyre-Walker. Understanding the degradation of hominid gene control. *PLoS Comput. Biol.* **2**, e19 (2006).

[19] M. Kimura. Evolutionary rate at the molecular level. *Nature* **217**, 624–626 (1968).

[20] M. Kimura. Preponderance of synonymous changes as evidence for the neutral theory of molecular evolution. *Nature* **267**, 275–276 (1977).

[21] T. I. Lee and R. A. Young. Transcription of eukaryotic protein-coding genes. *Annu. Rev. Genet.* **34**, 77–137 (2000).

[22] M. J. Lercher and L. D. Hurst. Human SNP variability and mutation rate are higher in regions of high recombination. *Trends Genet.* **18**, 337–340 (2002).

[23] L. A. Lettice, T. Horikoshi, S. J. Heaney, M. J. van Baren, H. C. van der Linde *et al.* Disruption of a long-range cis-acting regulator for Shh causes preaxial polydactyly. *Proc. Natl. Acad. Sci. USA* **99**, 7548–7553 (2002).

[24] H. Liang, Y. S. Lin and W. H. Li. Fast evolution of core promoters in primate genomes. *Mol. Biol. Evol. Mar.* 25 epub (2008).

[25] J. H. McDonald and M. Kreitman. Adaptive protein evolution at the Adh locus in Drosophila. *Nature* **351**, 652–654 (1991).

[26] F. Müller and L. Tora, The multicoloured world of promoter recognition complexes. *EMBO J.* **23**, 2–8 (2004).

[27] S. Myers L. Bottolo, C. Freeman, G. McVean and P. Donnelly. A fine-scale map of recombination rates and hotspots across the human genome. *Science* **310**, 321–324 (2005).

[28] T. D. Petes. Meiotic recombination hot spots and cold spots. *Nat. Rev. Genet.* **2**, 360–369 (2001).

[29] G. Piganeau and A. Eyre-Walker. Estimating the distribution of fitness effects from DNA sequence data: Implications for the molecular clock.*Proc. Natl. Acad. Sci. USA* **100**, 10335–10340 (2003).

[30] D. A. Pollard, V. N. Iyer, Moses and M. B. Eisen . Widespread discordance of gene trees with species tree in Drosophila: evidence for incomplete lineage sorting. *PLOS Genet.* **2**, e173 (2006).

[31] J. G. Prendergast, H. Campbell, N. Gilbert, M. G. Dunlop, W. A. Bickmore and C. A. Semple. Chromatin structure and evolution in the human genome. *BMC Evol. Biol.* **7**,72 (2007).

[32] M. V. Rockman, M. W. Hahn, N. Soranzo, F. Zimprich, D. B. Goldstein and G. A. Wray Ancient and recent positive selection transformed opioid cis-regulation in humans. *PLoS Biol.* **3**. e387 (2005).

[33] A. Sandelin, P. Carninci, B. Lenhard, J. Ponjavic, Y. Hayashizaki and D. A. Hume. Mammalian RNA polymerase II core promoters: insights from genome-wide studies. *Nat. Rev. Genet.* **8**, 424–436 (2007).

[34] C. A. Semple. Chromatin structure and human genome evolution. In *Encyclopedia of Life Sciences*. John Wiley & Sons Ltd. London. (2008).

[35] S. A. Shabalina, A. Y. Ogurtsov, V. A. Kondrashov and A. S. Kondrashov. Selective constraint in intergenic regions of human and mouse genomes. *Trends Genet.* **17**, 373–376 (2001).

[36] S. Shifman, J. T. Bell, R. R. Copley, M. S. Taylor, R. W. Williams, R. Mott and J. Flint. A high-resolution single nucleotide polymorphism genetic map of the mouse *genome. PLOS Biol.* **4**, e395 (2006).

[37] S. T. Smale, S. T. Kadonaga. The RNA polymerase II core promoter. *Annu. Rev. Biochem.* **72**, 449–479 (2003).

[38] B. E. Stranger, A. C. Nica, M. S. Forrest, A. Dimas, C. P. Bird, C. Beazley, C. E. Ingle, M. Dunning P. Flicek, D. Koller, S. Montgomery, S. Tavaré, P. Deloukas, E. T. Dermitzakis. Population genomics of human gene expression. *Nat. Genet.* **39**, 1217–1224 (2007).

[39] M. S. Taylor, C. Kai, J. Kawai, P. Carninci, Y. Hayashizaki and C. A. Semple Heterotachy in mammalian promoter evolution. *PLoS Genet.* **2**, e30 (2006).

[40] N. D. Trinklein, S. J. Aldred, A. J. Saldanha, R. M. Myers. Identification and functional analysis of human transcriptional promoters. *Genome Res.* **13**, 308–312 (2003).

[41] R. H. Waterston, K. Lindblad-Toh, E. Birney, J. Rogers, J. F. Abril, *et al.* Initial sequencing and comparative analysis of the mouse genome. *Nature* **420**, 520–562 (2002).

[42] C. L. Woodcock. Chromatin architecture. *Curr. Opin. Struct. Biol.* **16**, 213–220 (2006).

[43] G. A. Wray The evolutionary significance of cis-regulatory muta-tions. *Nat. Rev. Genet.* **8**, 206–216 (2007).

[44] M. C. King and A. C. Wilson. Evolution at two levels in humans and chimpanzees. *Science* **188**, (4184) 107–116 (1975).

[45] D. Farre, N. Bellora, L. Mularoni, X. Messeguer and M. M. Alba. Housekeeping genes tend to show reduced upstream sequence con-servation. *Genome Biol.* **8**(7):R140 (2007).

[46] M. S. Taylor and R. R. Copley. Comparative genomics. In *Bioinfor-matics for Geneticists* (2nd edition) (edited by Michael R. Barnes) John Wiley and Sons (2007).

[47] M. S. Taylor, T. Massingham, P. Carninci, Y. Hayashizaki, N. Gold-man and C. A. M. Semple. Rapidly evolving human promoter re-gions. *Nat. Genet.* **40**(11), 1262–1263 (2008).

Chapter Seventeen

The Impact of CAGE Data on Understanding Macrophage Transcriptional Biology

David A. Hume[1], Kate Schroder[2] and
Katharine M. Irvine[2,*]

[1]*The Roslin Institute and Royal (Dick) School of Veterinary Studies,
University of Edinburgh, UK*
[2]*Institute for Molecular Bioscience, University of Queensland,
Australia*
*Email: *khatarine.irvine@gmail.com*

This chapter demonstrates the relevance of CAGE data to studying promoter biology in depth. In particular, we report here what we have learned by from CAGE data obtained from macrophages, which are critical cellular mediators of innate immunity. Understanding transcriptional regulation and gene regulatory networks underlying basic biological mechanisms in these cells is key to understanding their physiological and pathological roles. The knowledge and methodological approaches to understanding macrophage biology outlined in this chapter can be applied to cells and tissues in diverse biological systems.

17.1 INTRODUCTION

Macrophages are the first line of defence against pathogens. They also mediate much of the pathology of infectious, inflammatory and malignant disease. The challenge in modulating macrophage action for therapeutic purposes is to balance these diverse macrophage functions; this requires a detailed understanding

*Cap Analysis Gene Expression (CAGE): The Science of Decoding Gene
Transcription* **edited by P Carninci**
Copyright © 2010 by Pan Stanford Publishing Pte Ltd
www.panstanford.com
978-981-4241-34-2

of gene regulation by factors that regulate macrophage differentiation and acquisition of effector functions ('macrophage activation'). Macrophages are part of the mononuclear phagocyte system, a cell lineage that includes precursors in bone marrow and circulating monocytes. Tissue macrophages comprise 10-15% of the cells in most organs of the body. They have roles in immunity and inflammation and are required for many aspects of development and normal homeostasis.[1−3] The production of macrophages from bone marrow progenitor cells is controlled by the major macrophage growth and differentiation factor, macrophage colony-stimulating factor, CSF-1 which signals through the CSF-1 receptor (CSF-1R).[4−6]

The regulation and function of the CSF-1 receptor is central to the biology of mononuclear phagocytes, as CSF-1R signalling controls many aspects of macrophage survival, proliferation and function.[4−6] Expression of *csf1r* mRNA is restricted to myeloid cells and their precursors.[3,7−9] During myeloid differentiation, the expression of *csf1r* mRNA is amongst the earliest events in macrophage lineage commitment, and the level of expression continues to increase as macrophages mature to monocytes and tissue macrophages.[10] Like many other macrophage-expressed genes, the *csf1r* gene is regulated by promoter elements for myeloid-restricted transcription factors such as PU.1 that are required for the production of myeloid cells from bone marrow progenitors (Reviewed in Reference 4).

CAGE is a technology that systematically identifies active promoters within a cellular system on a genome-wide scale, by sequencing the 5′ ends of capped transcripts (see Chapter 1 of this book). If sequencing is carried out to sufficient depth, CAGE tag frequency at a genomic location can be compared between two samples to indicate relative promoter use. The most extensive CAGE tag data to date was sequenced by the FANTOM (Functional Annotation of Mammals) consortium, which sequenced more than 7 million mouse, and 5 million human cage tags during FANTOM3.[11,12] An even greater depth of sequencing was achieved in FANTOM4, particularly in a major study of the differentiation of the human monocytic cell line, THP-1. Since macrophages are amongst the most complex and diverse sources of mRNA available in mammalian systems[13] the FANTOM3 consortium sequenced a substantial number of CAGE libraries derived from mouse macrophages responding to factors relevant to cell lineage commitment (CSF-1) and functions during infection (the bacterial cell wall component, lipopolysaccharide, LPS; and the macrophage-activating cytokine, interferon-gamma, IFN γ). Clustering these CAGE libraries based upon CAGE tag frequency generated two substantial promoter clusters that corresponded to macrophage-specific/enriched, and LPS-inducible sets.[12] This data provides a rich resource for ongoing studies of macrophage biology, at both a single gene level and genome-wide scale. It is also provides the opportunity for comparative studies with the large datasets currently emerging using human systems (e.g. FANTOM4 datasets). This chapter will review the impact of CAGE-derived data on our understanding of transcriptional biology in general, and macrophage biology in particular, and highlight the utility of this data for macrophage biologists.

17.2 TRANSCRIPTION START SITE AND PROMOTER CHARACTERISTICS REVEALED BY CAGE

CAGE tag and cDNA sequencing by the FANTOM3 consortium revealed a number of unexpected findings that greatly influence our understanding of the transcriptional landscape, including the finding that 63% of the genome is transcribed on at least one strand and the prevalence of mRNA-like non-coding transcripts.[11] Another unexpected discovery in this project was that the majority of promoters do not have a single defined transcription start site (so called "sharp" promoters). Instead, most genes initiate transcription from a cluster of nucleotide positions, spread over an area of 50-100 bp.[11,12] These promoter types have been classified as 'broad' promoters. The proportion of sharp promoters was somewhat under-estimated because their coverage is compromised by so-called multi-mapping; a subset of these sharp promoters share precise sequence identity in the proximal promoter region with other promoters in the genome.[14] But the fundamental observation remains. Interestingly, promoter classifications tend to be shared between orthologous mouse and human promoters.[15] Using CAGE technology, the FANTOM3 project identified 184–379 human and 177–349 mouse distinct TSS/TSS clusters.[12]

Another surprising finding was that most protein-coding genes do not conform to the conventional model whereby a gene has a single promoter driving transcription of a single mRNA species encoding a single protein product. In fact, most protein-coding genes employ more than one distinct promoter, shown by non-overlapping CAGE tag clusters, in some cases closely spaced, in other cases separated by more than 10 kb. Alternative promoters may permit independent regulation between different cell types, or in some cases, actually encode alternative N termini that change the function of the protein product.[16] Alternative promoter use is clearly evident in many genes expressed in macrophages.[17] In the recent study of human THP-1 cell differentiation, around half the protein-coding genes expressed in these cells used alternative promoters in this cell alone, and in many cases the alternative promoters were differentially regulated. In some cases, the CAGE data confirmed information obtained by more conventional approaches. For example, the data confirmed the presence of at least three separate promoters in the tartrate-resistant acid phosphatase gene.[15]

One of the more striking examples of alternative promoter use in the FANTOM3 CAGE data is the gelsolin gene.[11] Gelsolin is a cytoplasmic actin filament capping and severing protein. Figure 17.1(a) shows the genome element view of this gene. There are three distinct promoters; the upstream promoter is strongly enriched in macrophage libraries (CAGE tag frequencies around 5-fold enriched), although it is detectable in most tissue libraries. Additionally, a heart-specific promoter directs expression of an alternative transcription start, giving rise to an alternative 5′ exon, and directing alternative translation initiation. The alternative protein form has a signal peptide, and produces plasma gelsolin, a protein implicated in protection against tissue injury, especially in the context of toxic shock.[19,20] Interestingly, despite the enrichment in macrophages, gelsolin null macrophages have relatively few phenotypic deficiencies, in part because of expression of other members of the gelsolin family.[21] One such member, capG, is highly expressed by macrophages and shows highly restricted expression.[21] Figure 17.1(b) shows that in the capG gene there are also alternative promoters. In this case though, each of the promoters is strongly macrophage enriched (CAGE tag frequencies more than 10-fold higher than any tissue or cell source), and they simply encode alternative 5′UTRs. The function of these alternative forms is not known.

A second example with significant biological implication, shown in Fig. 17.2(a), is the complex locus encoding the key growth factor, insulin-like growth factor 1 (IGF-1). The production of IGF-1 by the liver is regulated by growth hormone (GH), and this is a central mechanism of control of growth in mammals. However, a liver-specific knockout of the mouse IGF-1 locus

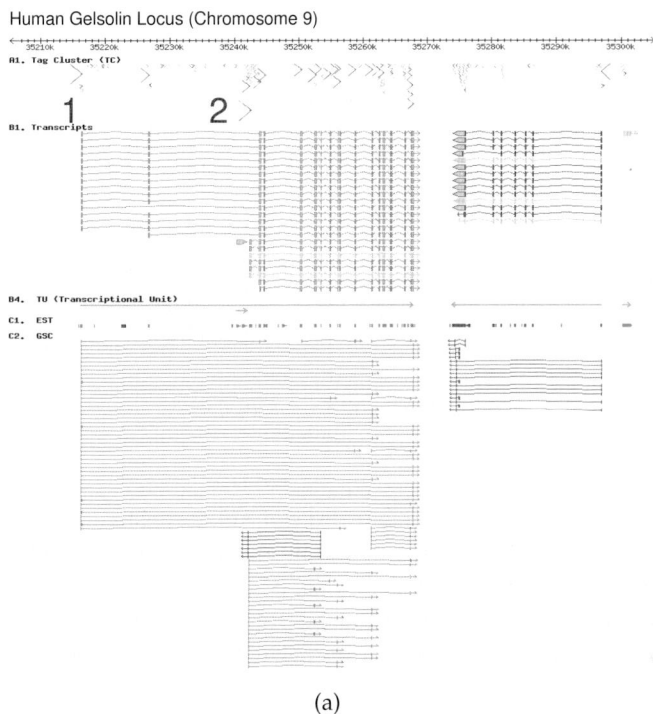

(a)

Figure 17.1. Genome element views of the mouse Gelsolin (a) and capG (b) loci. In each case, the panels show the tags clusters, full length cDNAs, and diTag sequences (from GSC, see Refs. 11 and 12) that identify paired 5′ and 3′ ends of transcripts. Note in panel a that there are two major promoters identified by the largest arrows. These generate alternative protein-coding N termini, promoter 1 is mainly used by macrophages, promoter 2 came mainly from lung libraries, but the tag frequency is highest in heart. Note also that the alternative promoters are supported by full length transcripts, and that gelsolin, but not the neighbouring gene, has significant numbers of tags over all the exons. In panel (b), the 5′ promoter is again largely macrophage-specific, and note that exonic promoter activity is largely absent.

Human CapG Locus (Chromosome 2)

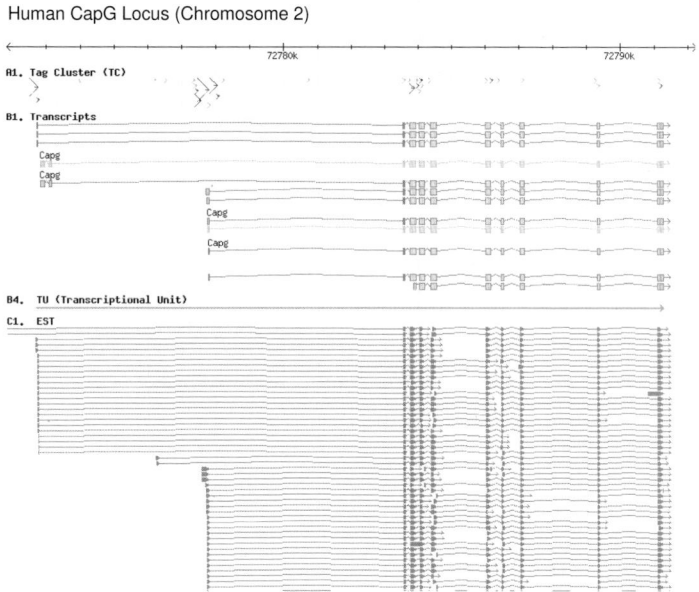

(b)

Figure 17.1 *Continued*

reduced circulating IGF-1 but did not ablate postnatal growth[22] The extensive CAGE data in the mouse reveal that there are at least 5 separable promoter regions, directing alternative 5′UTRs, within the IGF-1 locus. Amongst these, two TSS were identified almost exclusively from liver CAGE libraries, whilst others were very highly enriched in macrophage libraries. This finding reinforced earlier evidence that macrophages produce very high levels of *igf1* mRNA[23] and suggests they could be the major extrahepatic source. The most distal of the promoters is a TATA-containing promoter identified only in macrophage libraries, and only in the mouse. The CAGE tags allow a functional alignment of the IGF-1 promoter region across multiple mammalian species. Figure 17.2(b) shows the most highly-conserved region, bracketing a complex promoter used by both liver and macrophages. Figure 17.2(b) also shows alignment of the macrophage-specific TATA-containing promoter. This site was not detected in any human CAGE libraries, and indeed one can immediately see that the TATA box and PyPu initiator are both absent/mutated in non-rodents.

The FANTOM3 CAGE data was analysed for over-representation of promoter features amongst the "broad" and "sharp" TSS classes.[12] The TATA box, an element that tethers the pre-initiation complex to the DNA via its interaction with the TFIID component, TATA-binding protein (TBP), is known to have very precise positional preference with respect to the TSS. TATA-driven promoters comprised only a small fraction (10–20%) of promoters overall (as noted above, this may be something of an underestimate). As expected, the group of genes sharing sharp TSS were significantly correlated with the presence of a TATA box, presumably because the TATA box dictates the precise nucleotide for transcript initiation. This class of promoters was also associated with tissue- or context-specific gene expression. Nevertheless, many genes in this class of promoters did not contain a consensus TATA box, so there must be other elements that can specify a precise start site. Such evidence indicates that the textbook version of gene

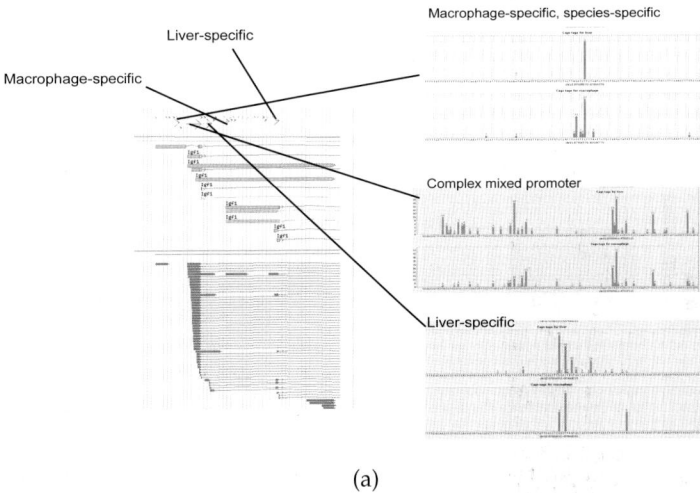

(a)

Figure 17.2. The mouse IGF-1 locus. Panel (a) shows the complex set of 5′ ends of the mouse IGF-1 locus. As in Fig. 17.1, these CAGE clusters are generally supported by full length cDNAs and by 5′ ESTs from full length cDNAs. The right hand segments show frequency histograms for particular CAGE tags mapped onto the genome, showing profiles for macrophages (top) and liver (bottom) in each case. Panel (b) shows a clustalW alignment of the macrophage-specific, and mixed promoters across mammalian species.

Macrophage-specific, rodent-specific, IGF-1 promoter region

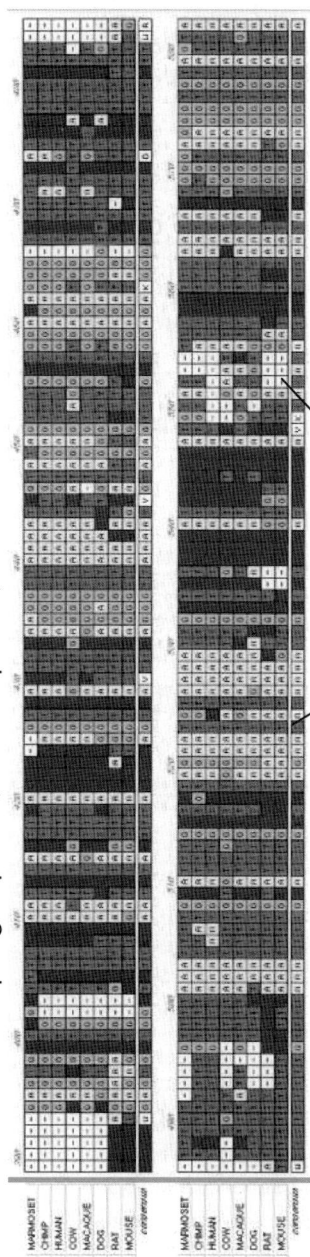

TATA Box and TSS are not conserved

Major liver and macrophage IGF-1 promoter region

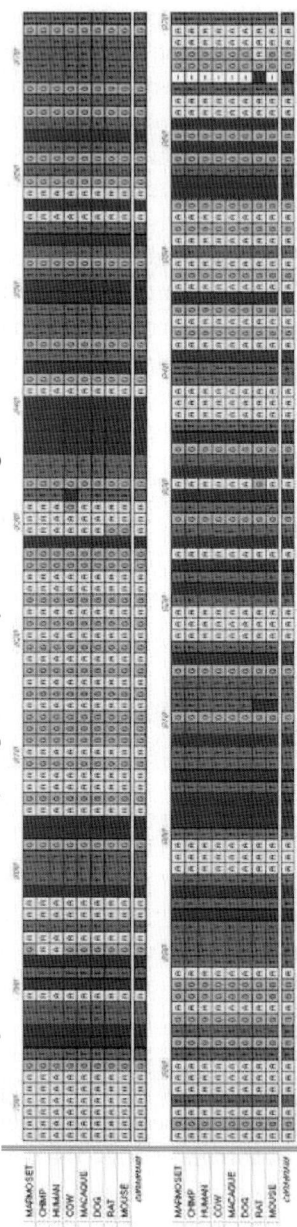

regulation in which a TATA-box directs transcript initiation at a precise nucleotide position is the exception rather than the rule.[15] The broad TSS promoters, on the other hand, were found to be strongly associated with CpG islands and expression of ubiquitously expressed genes.[15] TATA-containing promoters exhibited a lower substitution rate than TATA-less promoters, suggesting they evolve more slowly due to their constrained nature.[12]

Genome-wide sampling of transcription initiation sites at a single nucleotide resolution presents the opportunity to analyse nucleotide preferences for transcript initiation. This study suggested that the sequence requirements for transcript initiation were surprisingly permissive; strong conservation across the total set of core promoters was only found in the (-1, +1) dinucleotide, which had a preference for (pyrimidine, purine), and most promoters lacked the classical Inr consensus sequence.[12] This suggests that the TFIID complex actually binds to DNA relatively promiscuously, and then scans the DNA for a suitable TSS for RNA polymerase II-mediated transcription.[15,24]

As noted above, CAGE TSSs identified a cluster of core promoters corresponding to macrophage-enriched and -inducible transcripts, and these promoters were analysed for the presence of specific transcription factor-binding sites. The Ets and NFkB motifs were significantly over-represented within this promoter group, validating the utility of the approach as the Ets and NFkB transcription factor families are known to be major regulators of inflammatory and macrophage-specific gene expression.[12]

17.3 TRANSCRIPTIONAL COMPLEXITY: SENSE-ANTISENSE TRANSCRIPTION AND NON-CODING RNA

A key finding of the FANTOM3 collaboration was the prevalence of transcripts with no identifiable open reading frame (non-coding RNA). The function of the vast majority of these transcripts is yet to be defined. Many of these transcripts appear to participate in sense-antisense pairs, and experimental evidence suggested antisense transcription, at least in some cases, may exert regulatory effects on expression of the sense transcript.[25] We confirmed that a substantial subset of the non-coding mRNA-like transcripts are reproducibly expressed and regulated in macrophages, although clear evidence of sense-antisense regulation was lacking in most cases.[26] In fact, the overall prevalence is that sense-antisense pairs are more likely to be concordantly regulated.[25] A detailed analysis of such transcription suggested that there are ~1000 loci with at least three different transcripts in a sense-antisense relationship, or sharing a bi-directional promoter, that are conserved between human and mouse.[27] Such complexities are especially associated with genomic imprinting.[28] Non-coding mRNA-like transcripts tend to be expressed at significantly lower levels than protein-coding transcripts, and so were not as well covered by the original CAGE data. Nevertheless, although the transcripts themselves were not commonly conserved between mouse and human, the promoter regions were more conserved than those of protein-coding transcripts.[12] This finding suggests that the non-coding mRNA-like molecules function in cis, perhaps in a manner analogous to xist. It is now clear that the phenomenon of monoallelic inactivation, and allele counting, well known on the X chromosome, is shared with autosomes.[29,30] Interestingly, the LPS receptor encoded by the *tlr4* locus is monallelically expressed in macrophages.[30]

17.4 CONSTRUCTION OF MACROPHAGE TRANSCRIPTIONAL NETWORKS

In order to analyse the transcriptional networks operating in mammalian systems, promoter regions need to be defined and promoter activity measured to allow the prediction of transcription factor binding sites mediating gene regulation (Chapters 11 and 12). This can be done by overlaying microarray expression data with an arbitrary length of sequence upstream and downstream of the most 5′ nucleotide of a reference transcript (e.g. -1000 to +500 relative to the first nucleotide of a RefSeq). However this approach is imprecise, because the specific TSS in use in the cellular system of interest is not directly measured. A recent study[31] used CAGE-defined TSS to infer the transcriptional networks underlying macrophage activation by the archetypal inflammatory stimulus, LPS. This analysis indicated previously unappreciated key roles for the transcription factors ATF3 and NRF2; the importance of ATF3 in the LPS response was also predicted by another study.[32] Past efforts to construct transcriptional networks have often relied on microarray expression data to infer the activity of individual promoters. This approach compromises data quality because promoter activity is not measured directly, for example, many genes contain multiple promoters, making it impossible to accurately infer the relative activities of each promoter from (often 3') microarray data. If CAGE is performed in sufficient depth and replication (as was performed for FANTOM4), it allows the simultaneous measurement of exact TSS location and its relative expression dynamics, enabling higher quality predictions during network construction.

17.5 WHAT DOES CAGE DATA OFFER FOR TRADITIONAL STUDIES OF PROMOTER REGULATION?

The public CAGE data is an incredibly valuable resource for both genome-wide and more traditional single gene studies of transcriptional regulation. In order to allow the scientific community easy access to such data, we have created a portal on our web site at www.macrophages.com which allows gene-based admission to a range of informatic data from the public domain, including the FANTOM3 CAGE data, and our own cDNA microarray data on comparable samples. Described above were numerous examples of insights into the mechanics of transcription and transcriptional regulation enabled from CAGE-derived data, at a genome-wide scale. Here we will demonstrate the utility of CAGE to single gene projects, using the *csf1r* locus as an example.

In our earliest studies of the macrophage *csf1r* locus, we ascertained the transcription start site by both primer extension and RNase protection.[33] The mouse and human CAGE data provided a much more accurate annotation, and permitted a functional alignment across species. Figure 17.3 shows the CAGE data for the mouse and an alignment of the region upstream of the TSS cluster across multiple mammalian species. The first panel confirms that CAGE is entirely consistent with the previous RNA protection studies; there is a cluster of major start sites that confirm to the PyPu initiator consensus. Initiation of *csf1r* mRNA production in macrophages is controlled by a specific, broad class promoter that lacks the TATA-box, and also lacks the CpG islands or CCAAT box that are often characteristic of the broad class promoters.[4] The *csf1r* promoter contains multiple recognition sites for transcription factors of the Ets family, notably the lineage-restricted factor PU.1. This promoter architecture is shared with many other myeloid promoters, and is unique to the myeloid lineage.

In the region immediately upstream of the TSS cluster the mouse and rat promoters are clearly divergent from all other mammalian promoters. A limited number of studies of the activity of this promoter region, identified binding sites for PU.1, C/EBP family and AML-1 transcription factors. Interestingly, all three of these motifs are present in both the mouse and human promoters, but in distinct locations (Fig. 17.3). Two additional regulators that are likely to interact with the *csf1r* promoter are members of Fox and STAT families. Highly-conserved motifs conform to the consensus binding sites for these factors. In particular, FoxP1 may be a negative regulator of *csf1r* promoter activity.[34,35] FoxP1 is down-regulated as THP-1 or HL60 myeloid cells differentiate into macrophages in response to PMA, suggesting that differentiation is controlled in part by removal of repression. The published study did not examine the effect of mutating the FoxP1 element on promoter activity in macrophages, instead examining the ability of FoxP1 to block promoter activity in non-macrophage lines that do not normally express FoxP1 or *csf1r*. Similarly, the conserved site TTCxxxAGA resembles a STAT binding site. The differentiation of macrophages, in particular the early expression of *csf1r* in progenitors, is influenced by interleukin 3, GM-CSF and interferon-gamma[36,37] that act through different STAT factors.

The architecture of the *csf1r* promoter as a special class of mammalian promoter begs the question of precisely what defines the position of the TSS cluster. We hypothesized that a sequence immediately upstream of the TSS, in the region underlined in Fig. 17.3 may direct transcript initiation. We found that both the human and mouse sequences bound two proteins, FUS/TLS and Ewing sarcoma protein (EWSCR), that are both part of the basal transcription machinery.[38] We proposed that they recognise the transcription start site region in the absence of a TATA box, and that this may be a major mechanism to direct transcription amongst the TATA-less promoters. Unexpectedly, only EWSCR actually bound to this site in macrophage based upon ChIP analysis.[38] The two proteins also bind to G-rich sequences shared with the myeloid-zinc finger protein. This suggests that they could contribute to the G strand anisotropy seen in the more

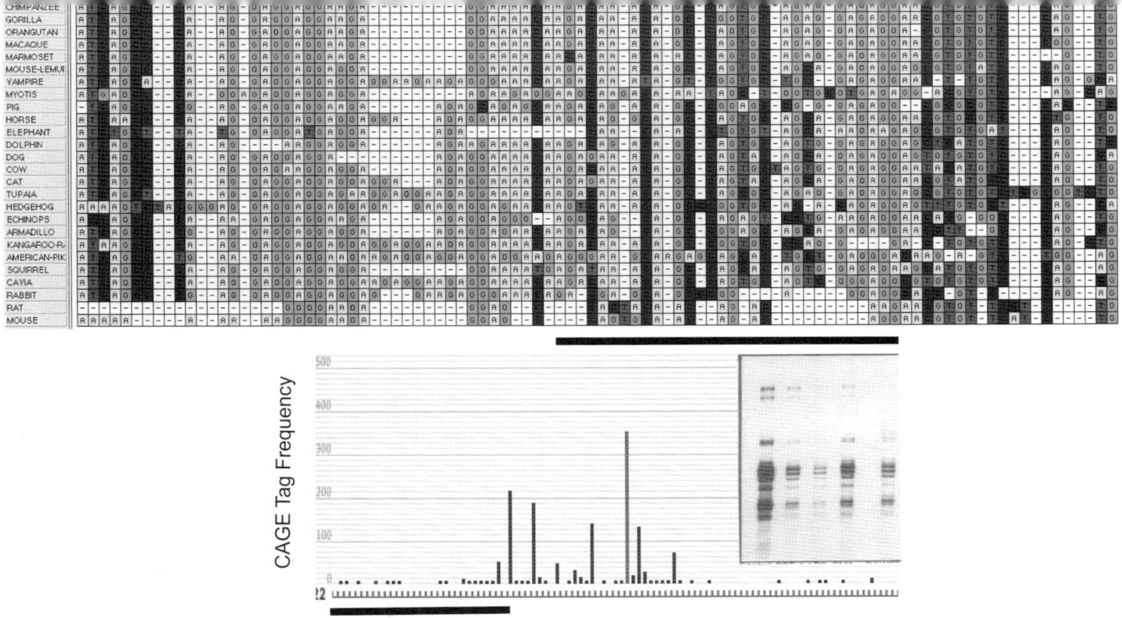

Figure 17.3. The mouse CSF-1 locus. The CAGE tag frequency histogram across the macrophage-specific *csf1r* promoter region (lower panel (http://www.macrophages.com)) confirms this is a broad type promoter, validating locus, showing a broad type promoting, and validating our much earlier TSS analysis using RNase protection (inset panel). The clustalW alignment in the upper part of the image shows that the region adjacent to the TSS cluster (underlined in both panels) is actually not highly conserved. Nevertheless, that sequence from both mouse and humans was shown to bind to the FUS and EWSCR proteins.

prevalent GC-rich class of "broad" promoters.[12] From the viewpoint of CAGE technology, the precise bp definition of the TSS cluster of this gene across both human and mouse systems was key to hypothesis generation and molecular studies of promoter function.

17.6 CONCLUSION

The development of an approach for simultaneous sequencing of CAGE tags from multiple libraries in a single reaction, allied to high-throughput sequencing[39] means that CAGE can now be viewed as a unique form of expression profiling. The innate immune system is under intense selection pressure to evolve in the face of selection by microorganisms. As many more mammalian genomes are completed, CAGE will provide a powerful tool for analysis of evolution of transcription initiation in macrophages.

References

[1] D. A. Hume. The mononuclear phagocyte system. *Curr. Opin. Immunol.* **18**, 49–53 (2006).

[2] D. A. Hume, The mononuclear phagocyte system revisited. *J. Leuk. Biol.* **72**, 621–627 (2002).

[3] F. Rae, Characterisation and trophic functions of murine embryonic macrophages based upon the use of a *csf1r*-EGFP transgene reporter. *Devt. Biol.* **308**, 232–246 (2007).

[4] C. Bonifer and D. A. Hume. The transcriptional regulation of the Colony-Stimulating Factor 1 Receptor (csf1r) gene during hematopoiesis. *Front Biosci.* **13**, 549–560 (2008).

[5] V. Chitu and E. R. Stanley. Colony-stimulating factor-1 in immunity and inflammation. *Current Opinion in Immunology* **18**, 39–48 (2006).

[6] M. J. Sweet and D. A. Hume. CSF-1 as a regulator of macrophage activation and immune responses. *Arch. Immunol Ther. Exp. (Warsz)* **51**, 169–177 (2003).

[7] A. M. Lichanska, C. M. Browne, G. W. Henkel, K. M. Murphy, M. C. Ostrowski, S. R. McKercher, R. A. Maki and D. A. Hume. Differentiation of the mononuclear phagocyte system during mouse embryogenesis: the role of transcription factor PU.1. *Blood* **94**, 127–138 (1999).

[8] A. M. Lichanska and D. A. Hume. Origins and functions of phagocytes in the embryo. *Exp. Hematol.* **28**, 601–611 (2000).

[9] R. T. Sasmono, A macrophage colony-stimulating factor receptor-green fluorescent protein transgene is expressed throughout the mononuclear phagocyte system of the mouse. *Blood* **101**, 1155–1163 (2003).

[10] H. Tagoh, R. Himes, D. Clarke, P. J. Leenen, A. D. Riggs, D. Hume and C. Bonifer. Transcription factor complex formation and chromatin fine structure alterations at the murine c-fms (CSF-1 receptor) locus during maturation of myeloid precursor cells. *Genes. Dev.* **16**, 1721–1737 (2002).

[11] P. Carninci *et al*. The transcriptional landscape of the mammalian genome. *Science* **309**, 1559–1563 (2005).

[12] P. Carninci *et al*. Genome-wide analysis of mammalian promoter architecture and evolution. *Nature Genet.* **38**, 626–635 (2006).

[13] C. A. Wells, T. Ravasi, R. Sultana, K. Yagi, P. Carninci, H. Bono, G. Faulkner, Y. Okazaki, J. Quackenbush, D. A. Hume, RIKEN GER Group, GSL Members and P. A. Lyons. Continued discovery of transcriptional units expressed in cells of the mouse mononuclear phagocyte lineage. *Genome Res.* **13**, 1360–1365 (2003).

[14] G. J. Faulkner, A. R. Forrest, A. M. Chalk, K. Schroder, Y. Hayashizaki, P. Carninci, D. A. Hume and S. M. Grimmond. A rescue strategy for multimapping short sequence tags refines surveys of transcriptional activity by CAGE. *Genomics* **91**, 281–288 (2008).

[15] A. Sandelin, P. Carninci, B. Lenhard, J. Ponjavic, Y. Hayashizaki, and D. A. Hume. Mammalian RNA polymerase II core promoters: Insights from genome-wide studies. *Nature Rev. Gene.* **8**, 424–436 (2007).

[16] M. J. Davis, K. A. Hanson, F. Clark, J. L. Fink, F. Zhang, T. Kasukawa, C. Kai, J. Kawai, P. Carninci, Y. Hayashizaki and R. D. Teasdale. Differential use of signal peptides and membrane domains is a common occurrence in the protein output of transcriptional units. *PLoS Genet.* **2**, e46 (2006).

[17] C. A. Wells, A. M. Chalk, A. Forrest, D. Taylor, N. Waddell, K. Schroder, S. R. Himes, G. Faulkner, S. Lo, T. Kasukawa, H. Kawaji, C. Kai, J. Kawai, S. Katayama, P. Carninci, Y. Hayashizaki, D. A. Hume and S. M. Grimmond. Alternate transcription of the Toll-like receptor signaling cascade. *Genome Biol.* **7**, R10 (2006).

[18] N. C. Walsh, M. Cahill, P. Carninci, J. Kawai, Y. Okazaki, Y. Hayashizaki, D. A. Hume and A. I. Cassady. Multiple tissue-specific promoters control expression of the murine tartrate-resistant acid phosphatase gene. *Gene* **307**, 111–123 (2003).

[19] P. S. Lee, A. B. Waxman, K. L. Cotich, S. W. Chung, M. A. Perrella and T. P. Stossel. Plasma gelsolin is a marker and therapeutic agent in animal sepsis. *Critical Care Medicine* **35**, 849–855 (2007).

[20] L. Spinardi and W. Witke. Gelsolin and diseases. *Sub-cellular Biochemistry* **45**, 55–69 (2007).

[21] W. Witke, W. Li, D. J. Kwiatkowski and F. S. Southwick. Comparisons of CapG and gelsolin-null macrophages: demonstration of a

unique role for CapG in receptor-mediated ruffling, phagocytosis, and vesicle rocketing. *J. Cell Biol.* **154**, 775–784 (2001).

[22] A. A. Butler and D. LeRoith. Minireview: Tissue-specific versus generalized gene targeting of the igf1 and igf1r genes and their roles in insulin-like growth factor physiology. *Endocrinology* **142**, 1685–1688 (2001).

[23] S. Arkins, N. Rebeiz, A. Biragyn, D. L. Reese and K. W. Kelley. Murine macrophages express abundant insulin-like growth factor-I class I Ea and Eb transcripts. *Endocrinology* **133**, 2334–2343 (1993).

[24] M. C. Frith, E. Valen, A. Krogh, Y. Hayashizaki, P. Carninci, and A. Sandelin. A code for transcription initiation in mammalian genomes. *Genome Res.* **18**, 1–12 (2008).

[25] S. Katayama *et al.* Antisense transcription in the mammalian transcriptome. *Science* **309**, 1564–1566 (2005).

[26] T. Ravasi, H. Suzuki, K. C. Pang, S. Katayama, M. Furuno, R. Okunishi, S. Fukuda, K. Ru, M. C. Frith, M. M. Gongora, S. M. Grimmond, D. A. Hume, Y. Hayashizaki and J. S. Mattick. Experimental validation of the regulated expression of large numbers of non-coding RNAs from the mouse genome. *Genome Res.* **16**, 11–19 (2006).

[27] P. G. Engstrom *et al.* Complex Loci in human and mouse genomes. *PLoS Genet.* **2**, e47 (2006).

[28] R. Holmes, C. Williamson, J. Peters, P. Denny and C. Wells. A comprehensive transcript map of the mouse Gnas imprinted complex. *Genome Research* **13**, 1410-1415 (2003).

[29] A. Gimelbrant, J. N. Hutchinson, B. R. Thompson and A. Chess. Widespread monoallelic expression on human autosomes. *Science* **318**, 1136–1140 (2007).

[30] N. Singh, F. A. Ebrahimi, A. A. Gimelbrant, A. W. Ensminger, M. R. Tackett, P. Qi, J. Gribnau and A. Chess. Coordination of the random asynchronous replication of autosomal loci. *Nat. Genet.* **33**, 339–341 (2003).

[31] R. Nilsson, V. B. Bajic, H. Suzuki, D. di Bernardo, J. Bjorkegren, S. Katayama, J. F. Reid, M. J. Sweet, M. Gariboldi, P. Carninci, Y. Hayashizaki, D. A. Hume, J. Tegner and T. Ravasi. Transcriptional network dynamics in macrophage activation. *Genomics* **88**, 133–142 (2006).

[32] M. Gilchrist, V. Thorsson, B. Li, A. G. Rust, M. Korb, K. Kennedy, T. Hai, H. Bolouri and A. Aderem. Systems biology approaches identify ATF3 as a negative regulator of Toll-like receptor 4. *Nature* **441**, 173–178 (2006).

[33] X. Yue, P. Favot, T. L. Dunn, A. I. Cassady and D. A. Hume. Expression of mRNA encoding the macrophage colony-stimulating factor receptor (c-fms) is controlled by a constitutive promoter and tissue-specific transcription elongation. *Mol. Cell. Biol.* **13**, 3191–3201 (1993).

[34] C. Shi and D. I. Simon. Integrin signals, transcription factors and monocyte differentiation. *Trends Cardiovasc. Med.* **16**, 146-152 (2006).

[35] C. Shi, X. Zhang, Z. Chen, K. Sulaiman, M. W. Feinberg, C. M. Ballantyne, M. K. Jain and D. I. Simon. Integrin engagement regulates monocyte differentiation through the forkhead transcription factor Foxp1. *J. Clin. Invest.* **114**, 408–418 (2004).

[36] F. N. Breen, D. A. Hume and M. J. Weidemann. The effects of interleukin 3 (IL-3) on cells responsive to macrophage colony-stimulating factor (CSF-1) in liquid murine bone marrow culture. *Br. J. Haematol.* **74**, 138–145 (1990).

[37] N. Breen, D. A. Hume and M. J. Weidemann. Interactions among granulocyte-macrophage colony-stimulating factor, macrophage colony-stimulating factor and IFN-gamma lead to enhanced proliferation of murine macrophage progenitor cells. *J. Immunol.* **147**, 1542–1547 (1991).

[38] D. A. Hume, T. Sasmono, S. R. Himes, S. M. Sharma, A. Bronitz, M. Constantin, M. C. Ostrowski and I. L. Ross. The Ewing sarcoma protein (EWS) binds directly to proximal promoter elements of the macrophaage-specific promoter of the CSF-1 receptor (c-fms) gene. *J. Immunol.* **180**, 6733–6742 (2008).

[39] N. Maeda, H. Nishiyori, M. Nakamura, C. Kawazu, M. Murata, H. Sano, K. Hayashida, S. Fukuda, M. Tagami, A. Hasegawa, K. Murakami, K. Schroder, K. Irvine, D. A. Hume, Y. Hayashizaki, P. Carninci and H. Suzuki. Development of a DNA barcode tagging method for monitoring dynamic changes in gene expression using ultra high throughput sequencer. *Biotechniques* **45**, 95–97 (2008).

Color Index

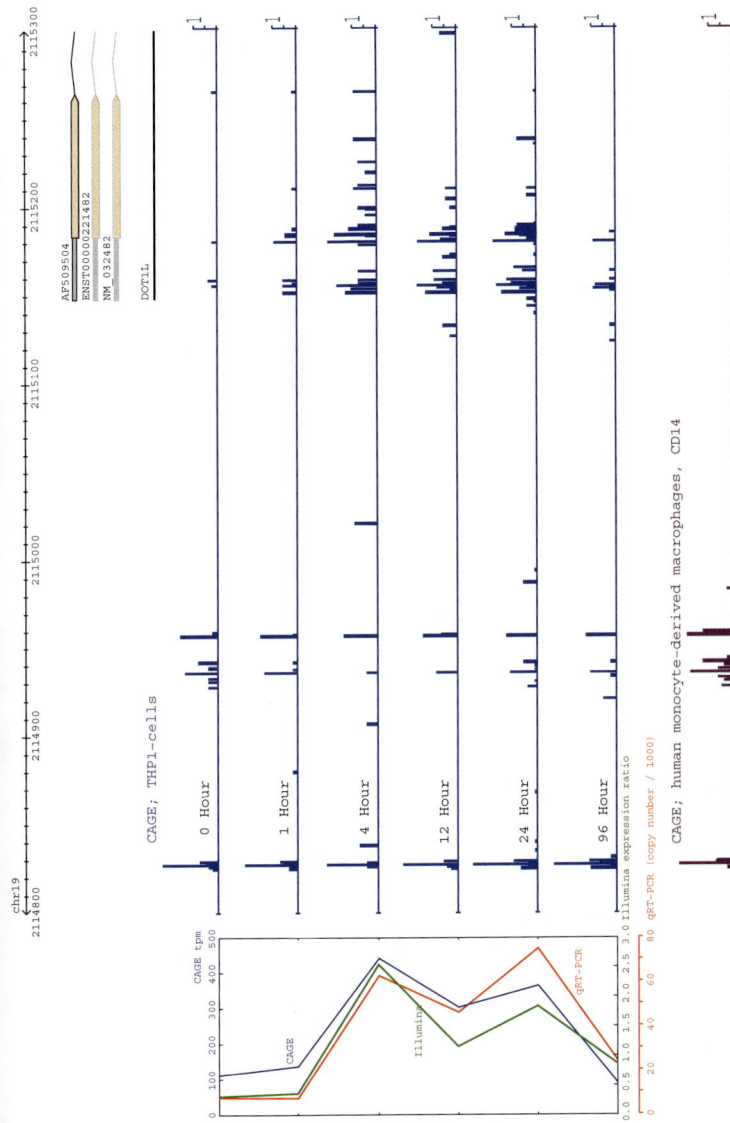

Figure 1.5 Extracted from Chapter 1, Page 5.

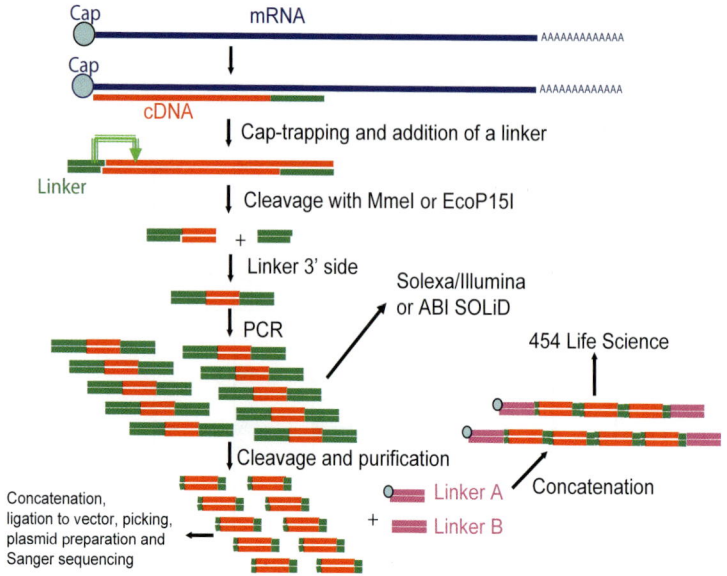

Figure 2.1   Extracted from Chapter 2, Page 12.

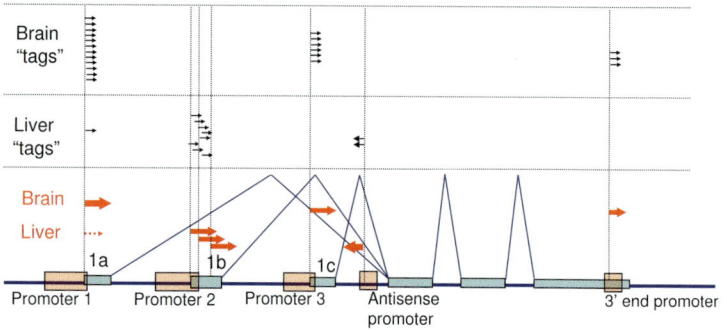

Figure 2.2   Extracted from Chapter 2, Page 15.

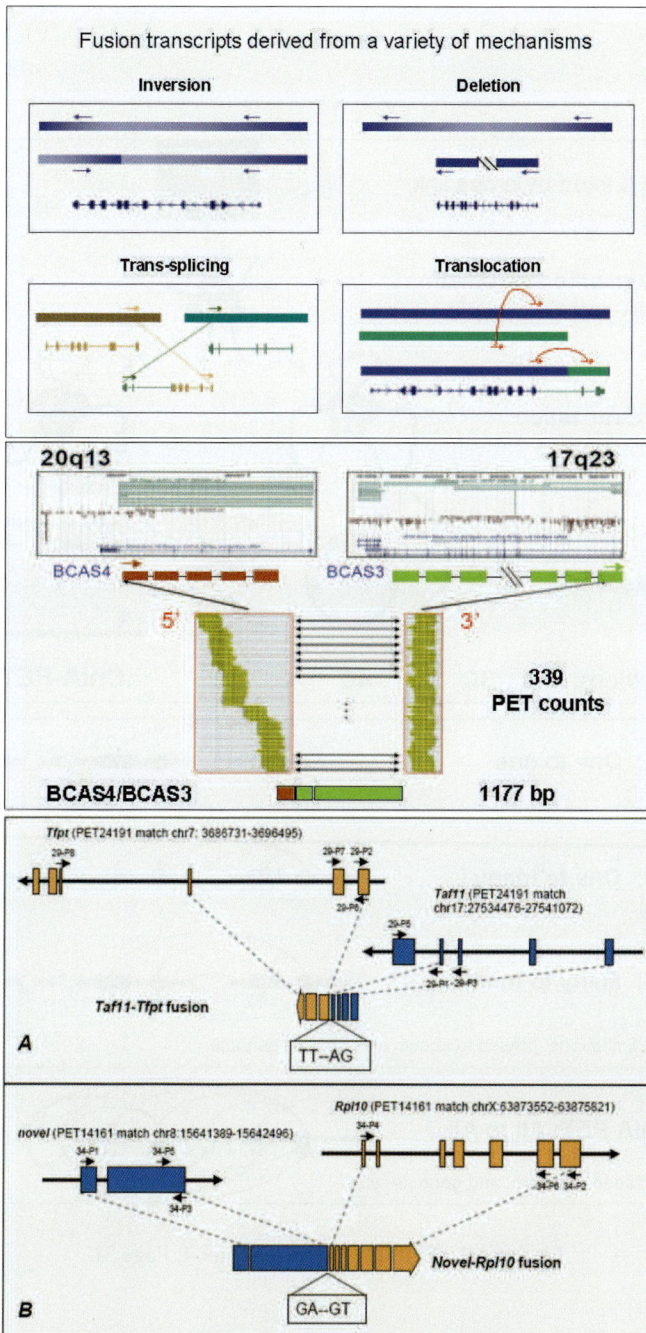

Figure 4.3 Extracted from Chapter 4, Page 49.

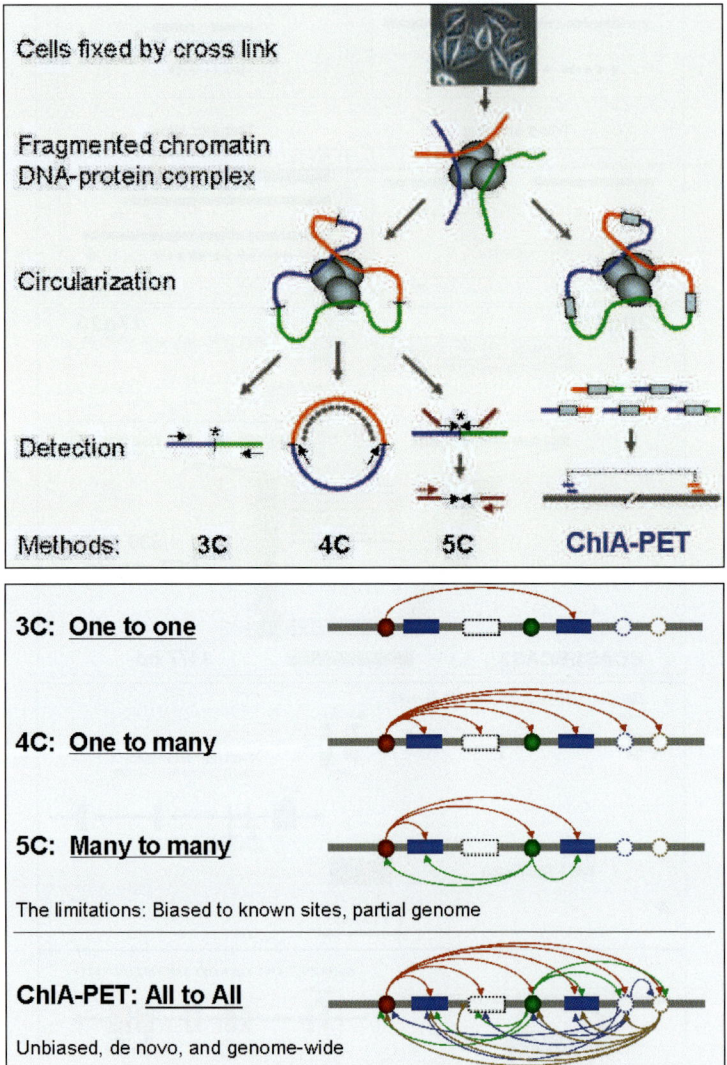

Figure 4.5 Extracted from Chapter 4, Page 54.

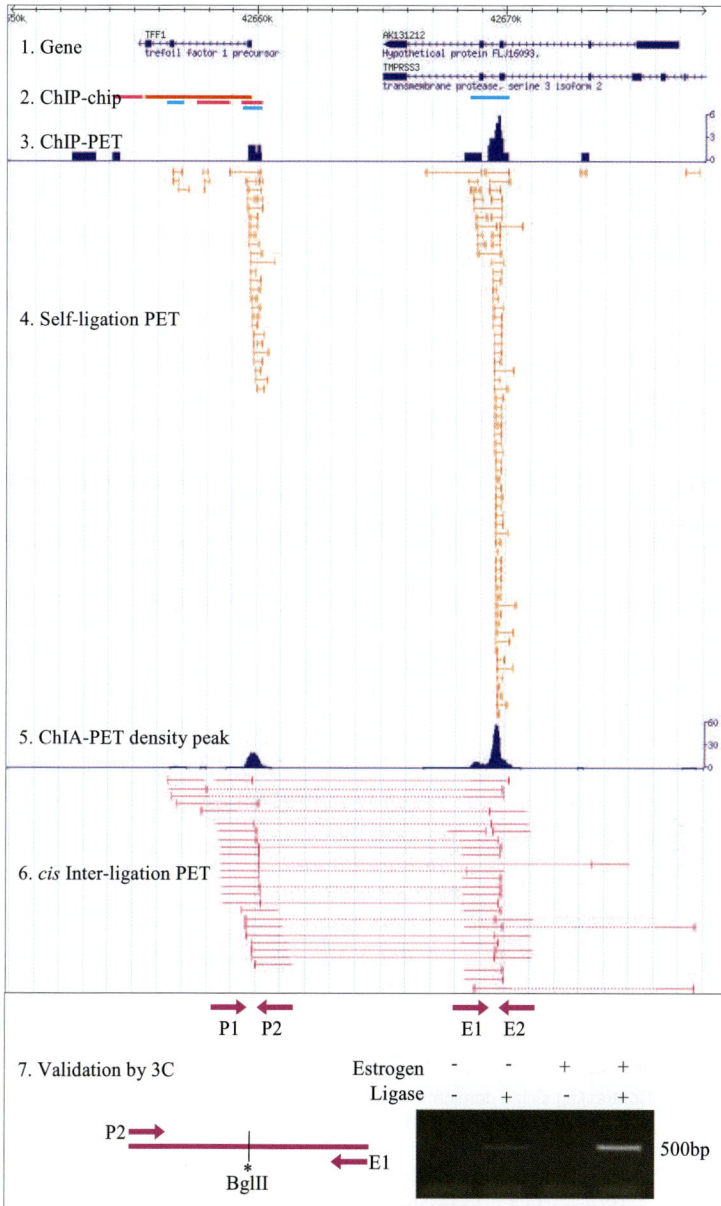

Figure 4.6   Extracted from Chapter 4, Page 56.

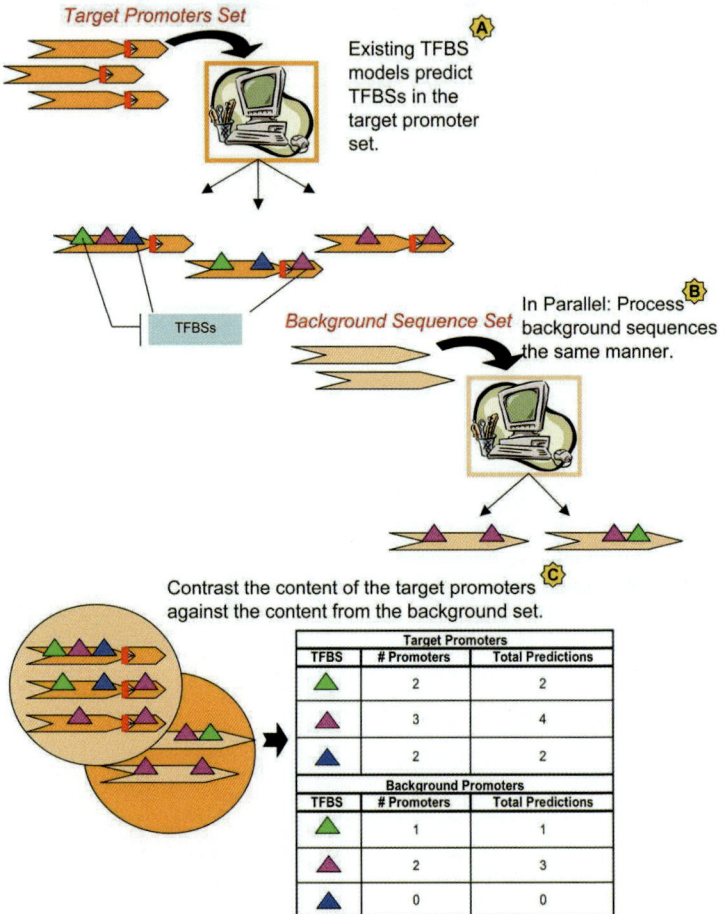

Figure 11.1 Extracted from Chapter 11, Page 140.

Phase 3: Refining TFBS predictions

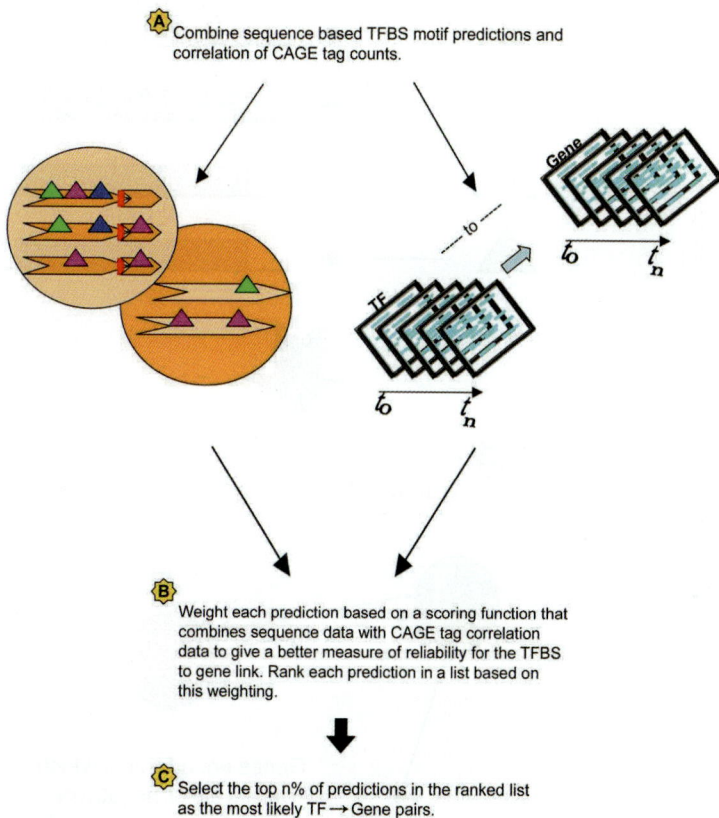

Ⓐ Combine sequence based TFBS motif predictions and correlation of CAGE tag counts.

Ⓑ Weight each prediction based on a scoring function that combines sequence data with CAGE tag correlation data to give a better measure of reliability for the TFBS to gene link. Rank each prediction in a list based on this weighting.

Ⓒ Select the top n% of predictions in the ranked list as the most likely TF → Gene pairs.

Figure 11.3 Extracted from Chapter 11, Page 142.

Phase 4: Constructing regulatory network

List of top ranked predictions.

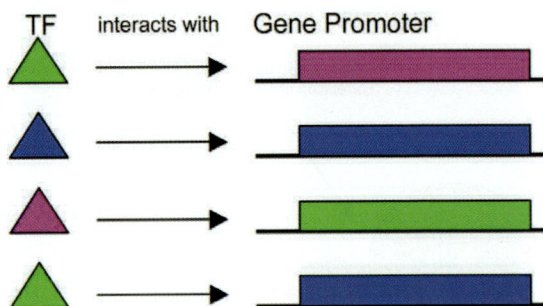

TF   interacts with   Gene Promoter

Transcriptional Regulatory Network

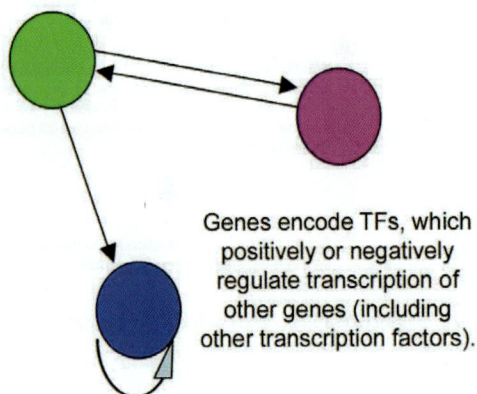

Genes encode TFs, which positively or negatively regulate transcription of other genes (including other transcription factors).

Figure 11.4   Extracted from Chapter 11, Page 143.

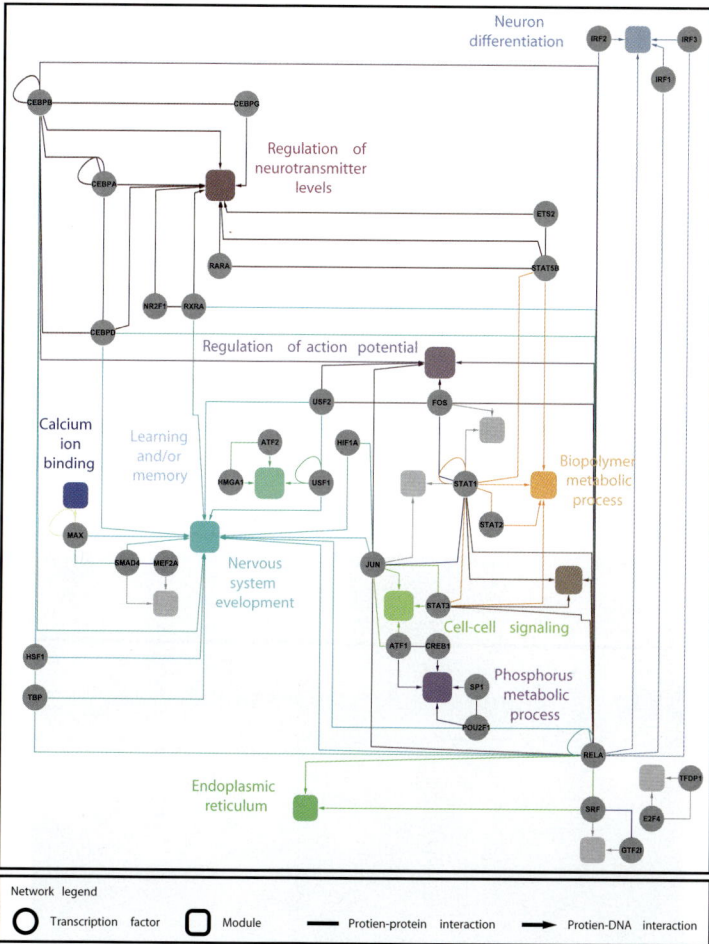

Figure 12.2 Extracted from Chapter 12, Page 159.

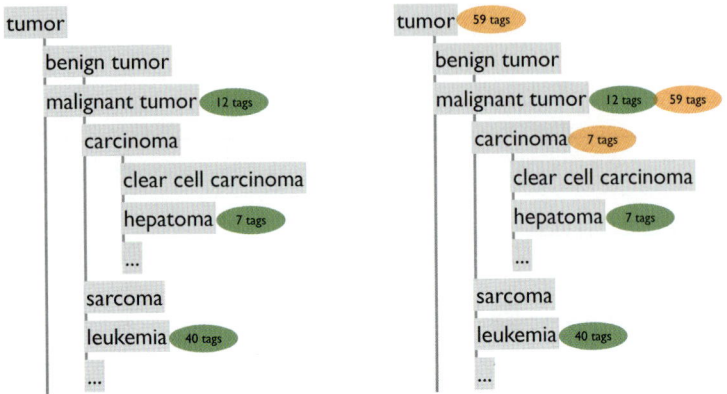

Figure 13.2 Extracted from Chapter 13, Page 175.

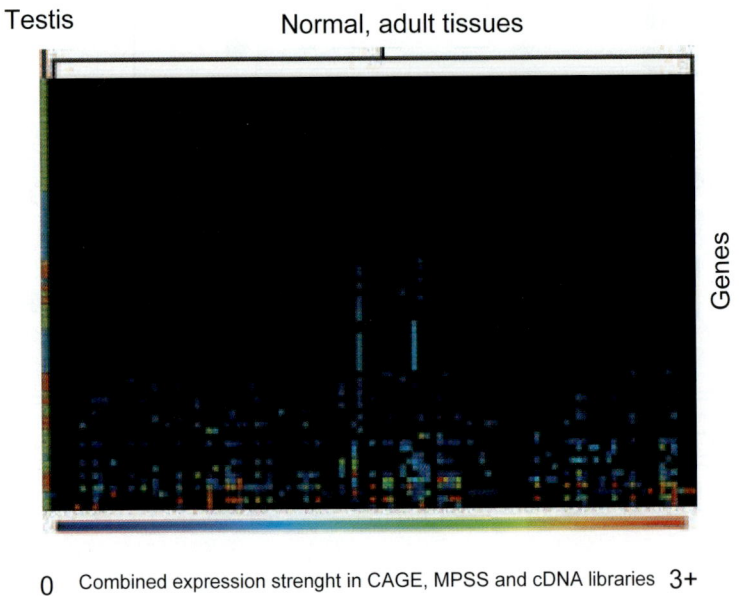

Testis   Normal, adult tissues

Genes

0   Combined expression strenght in CAGE, MPSS and cDNA libraries   3+

Figure 13.4 Extracted from Chapter 13, Page 177.

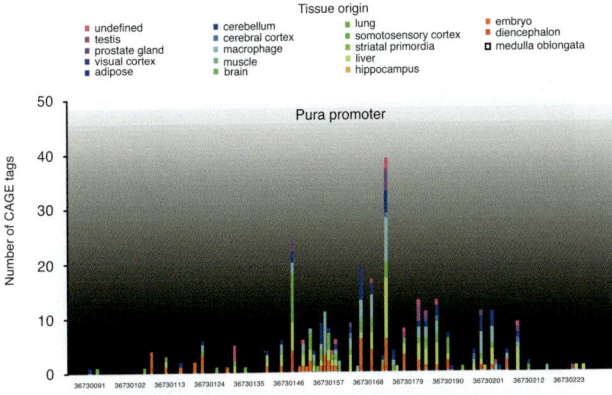

(a)

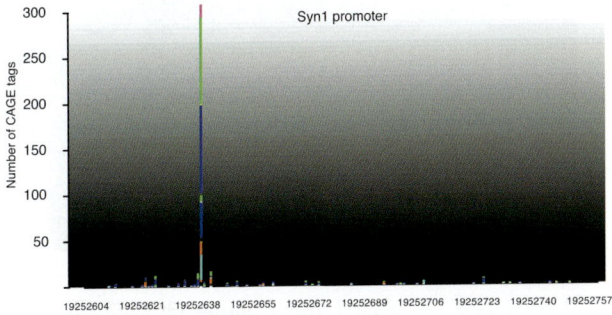

(b)

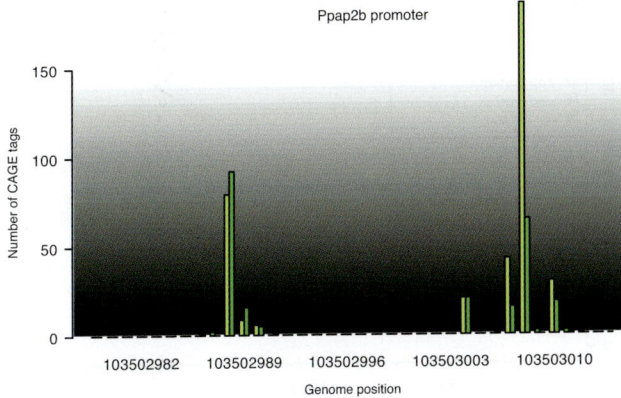

(c)

Figure 14.3 Extracted from Chapter 14, Page 185 and 186.

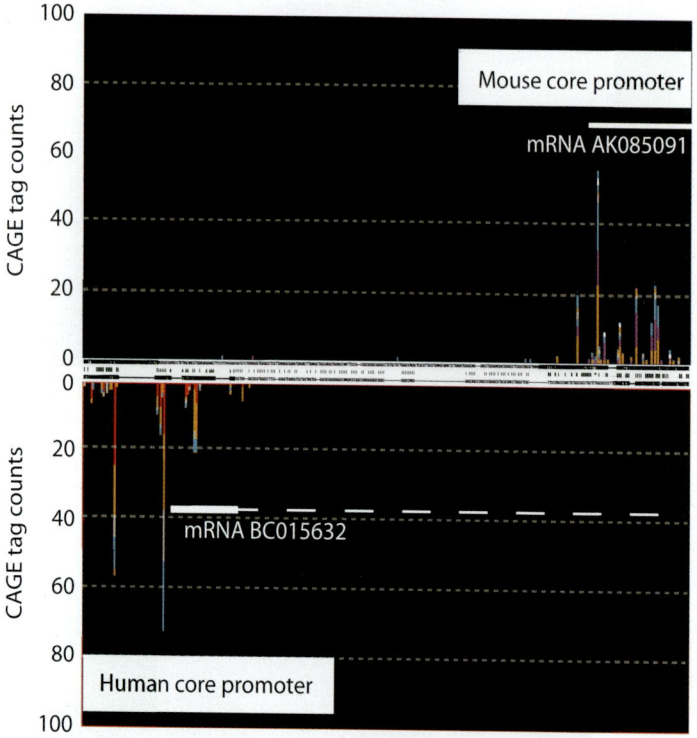

Figure 14.4   Extracted from Chapter 14, Page 193.

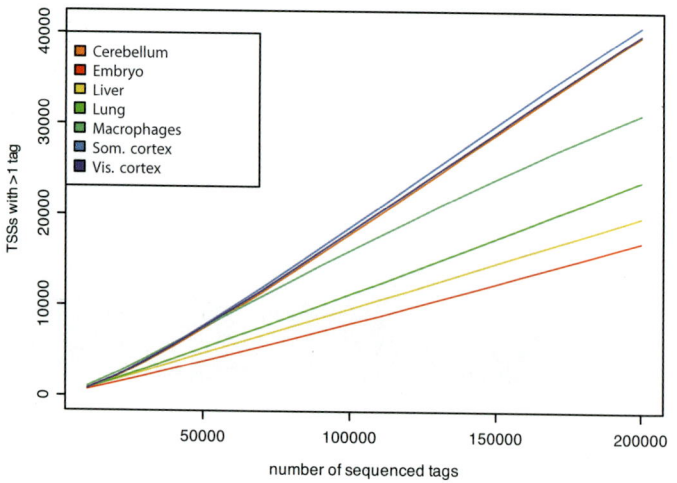

Figure 15.1   Extracted from Chapter 15, Page 200.

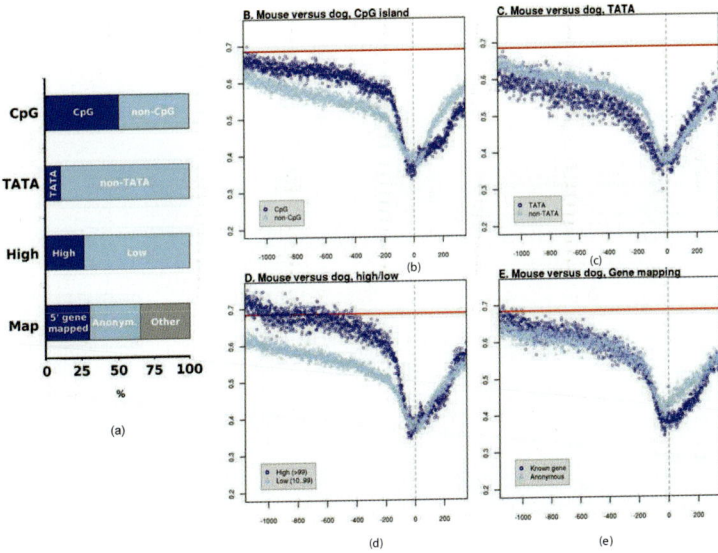

Figure 16.2   Extracted from Chapter 14, Page 216.

Human Gelsolin Locus (Chromosome 9)

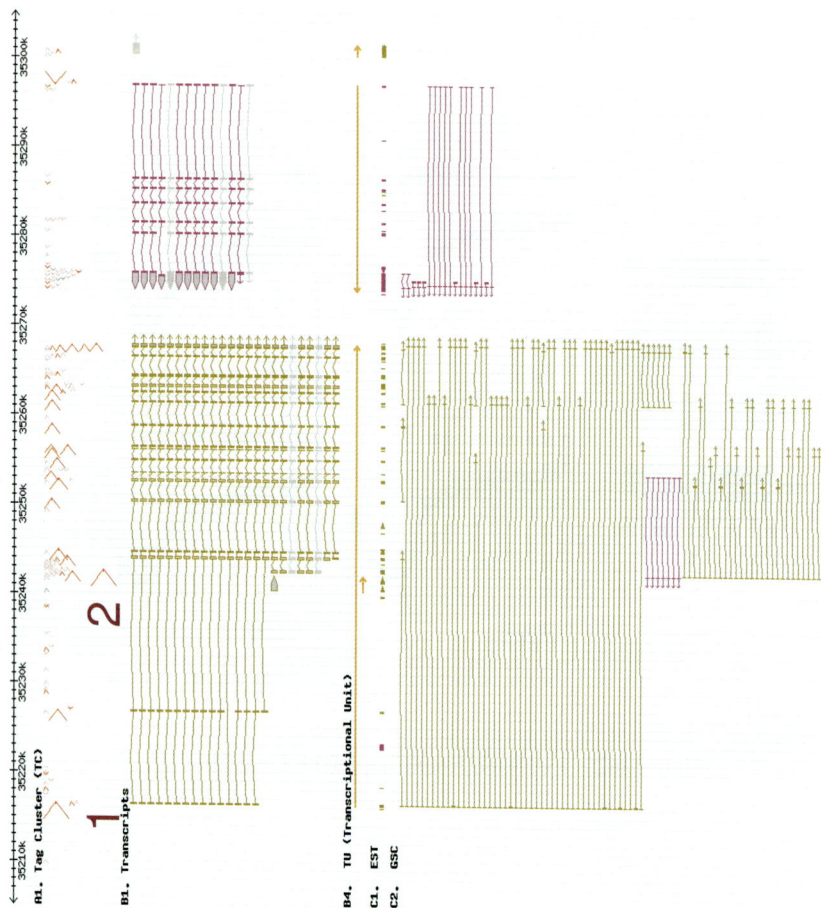

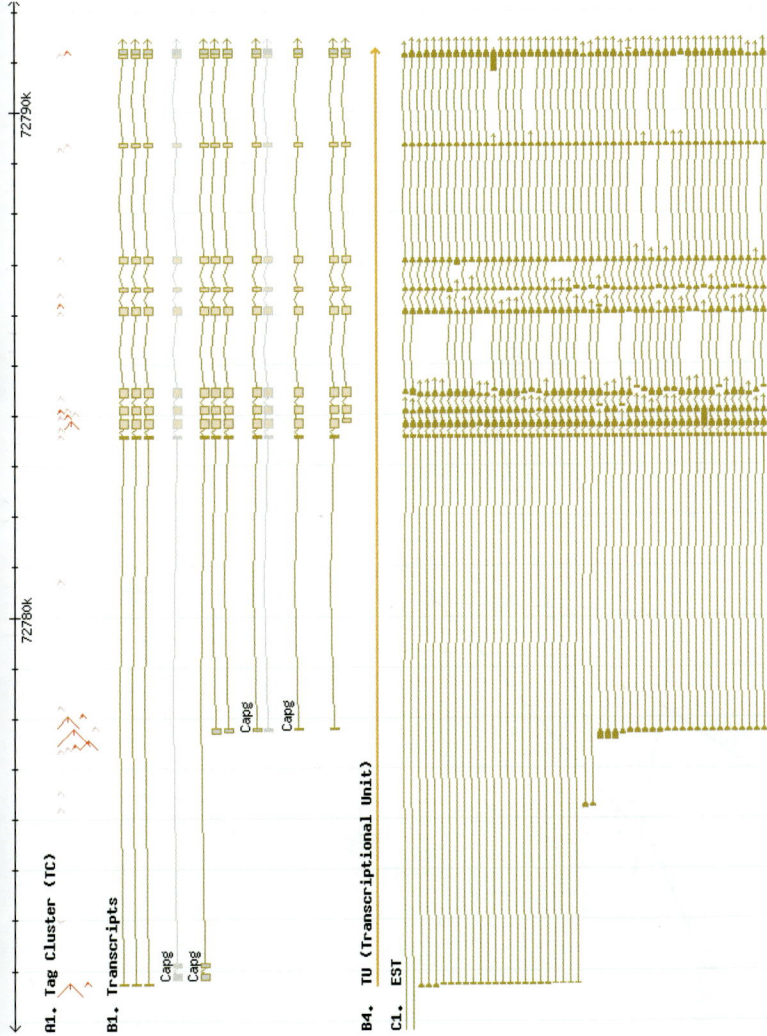

A1. Tag Cluster (TC)

B1. Transcripts

B4. TU (Transcriptional Unit)

C1. EST

Capg

72780k

72790k

(b)

Figure 17.1 *Continued* Extracted from Chapter 17, Page 232.

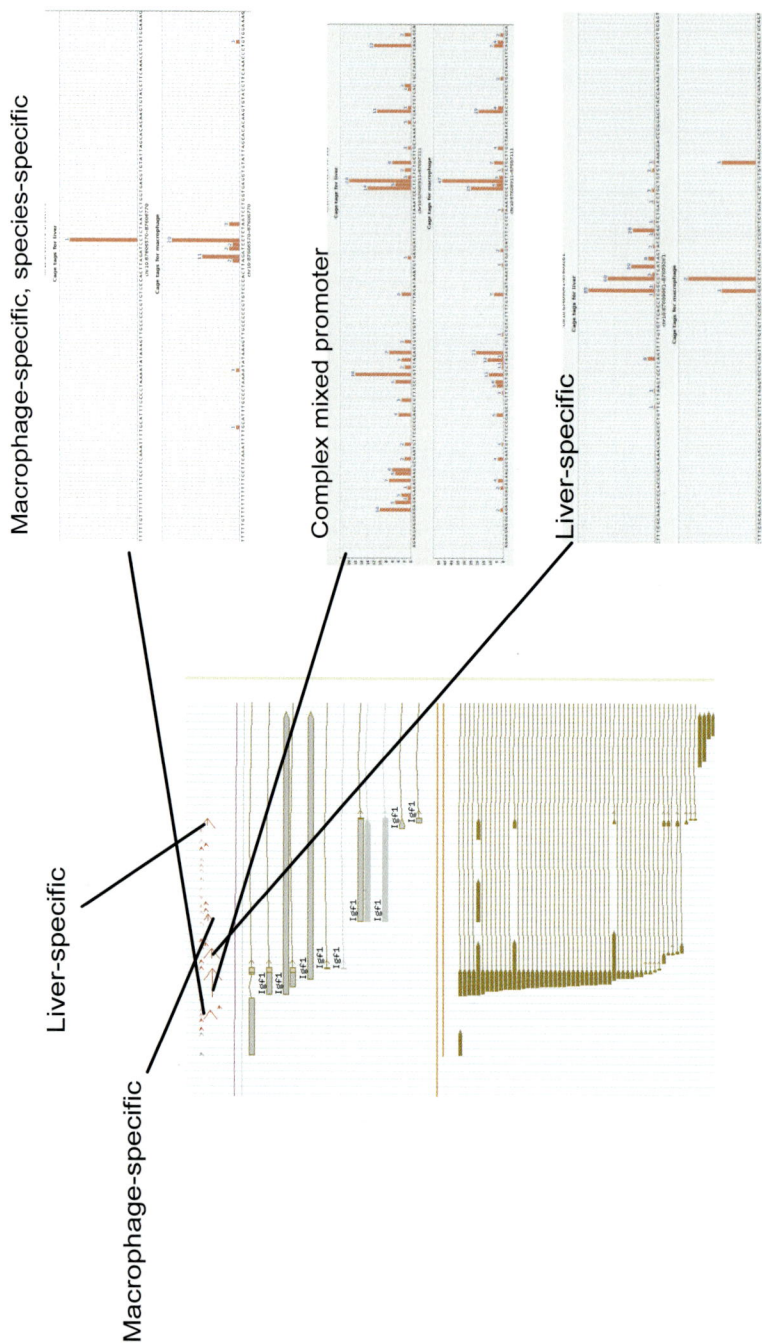

Macrophage-specific, species-specific

Complex mixed promoter

Liver-specific

Liver-specific

Macrophage-specific

(a)

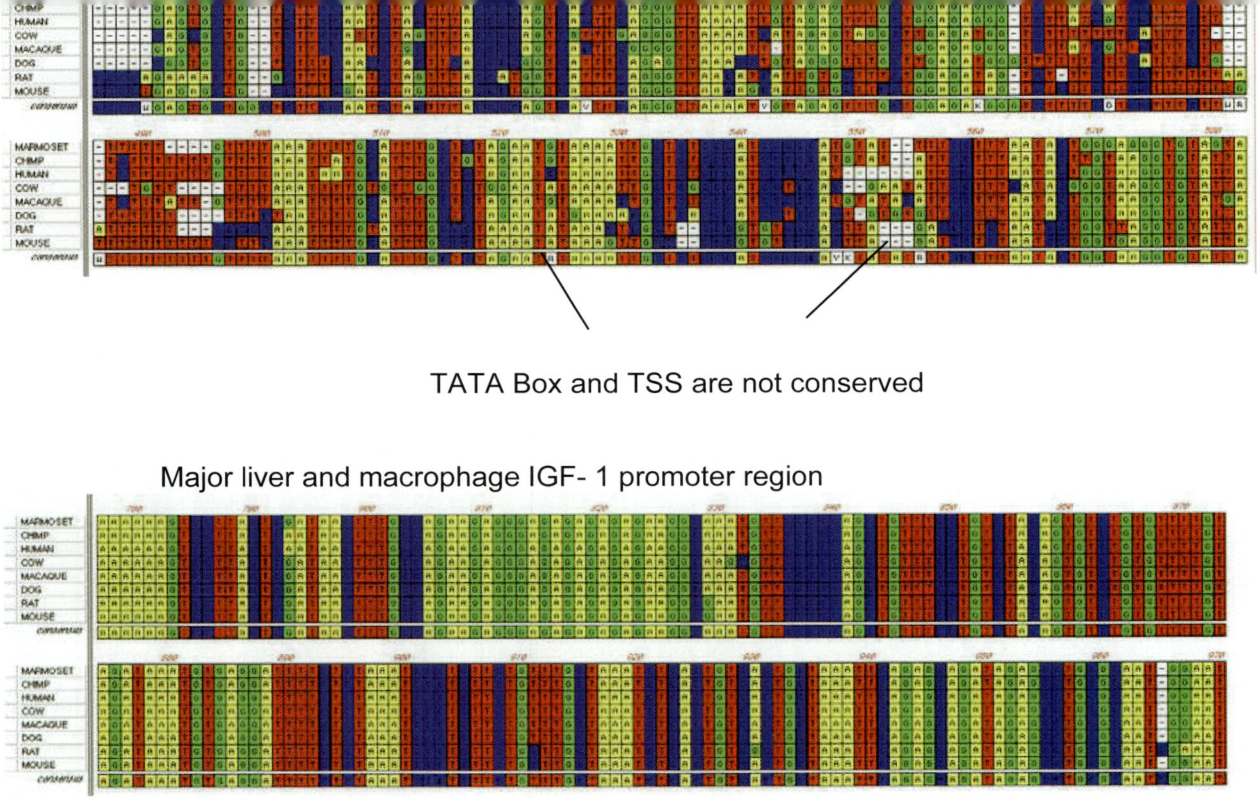
TATA Box and TSS are not conserved
Major liver and macrophage IGF- 1 promoter region
MARMOSET
CHIMP
HUMAN
COW
MACAQUE
DOG
RAT
MOUSE

(b)

Figure 17.2 *Continued* Extracted from Chapter 17, Page 234.

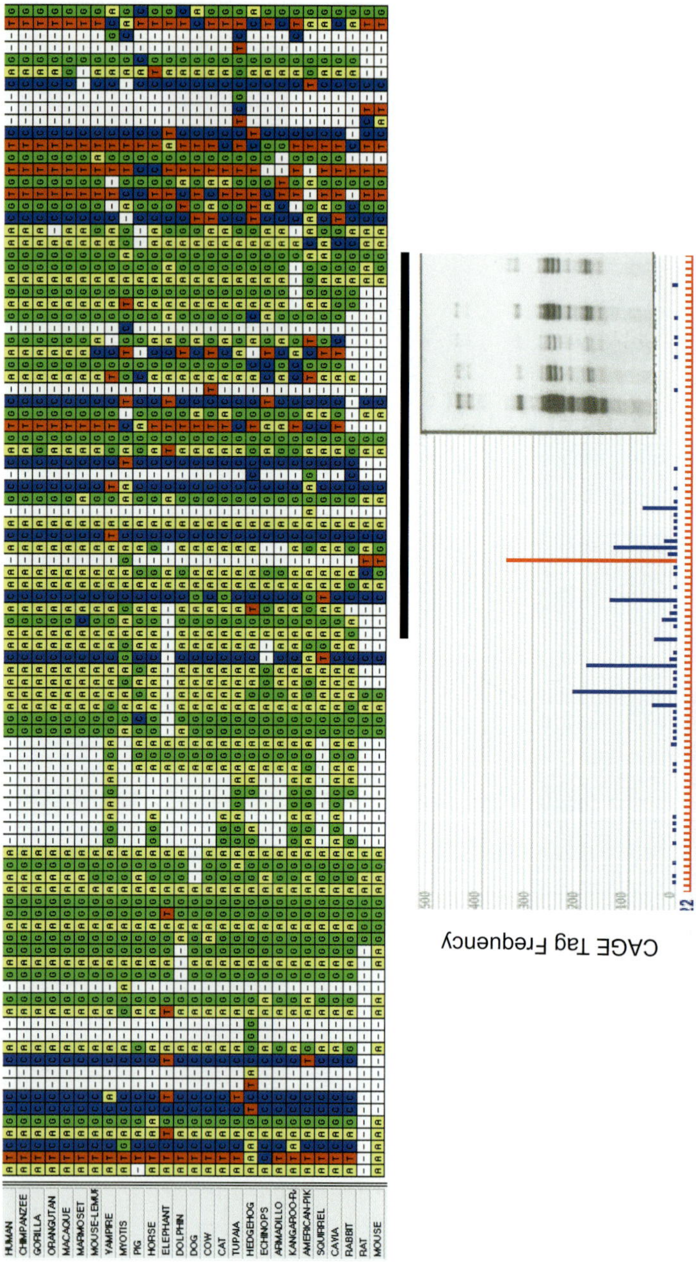

Figure 17.3 Extracted from Chapter 17, Page 239.

Index